BIODYNAMIC WINE

BIODYNAMIC
WINE

MONTY WALDIN

Monty Waldin sensed, when working as a teenager on a conventionally run Bordeaux château in the mid-1980s, that the more unnecessary sprays were applied to the grapes, the more additives and other corrective treatments were needed subsequently during the winemaking. By the mid-1990s Monty had become the first wine writer to specialize in green issues. His first book, *The Organic Wine Guide* (Thorsons, 1999), published whilst Monty developed a biodiversity project for a Demeter-certified biodynamic vineyard in California, was voted Britain's Wine Guide of the Year. He followed this with the multi-award-winning *Biodynamic Wines* (Mitchell Beazley, 2004). Monty drew on his winemaking experiences in Chile for *Wines of South America* (Mitchell Beazley, 2003), winner of America's prestigious James Beard Book Award. In 2007, whilst living in the Roussillon in France, Monty was filmed by Britain's Channel 4 for *Château Monty*, the first ever observational TV documentary on biodynamic winemaking from pruning to bottling (a six-part series broadcast in 2008). His other books include *Discovering Wine Country: Bordeaux* (2005, Mitchell Beazley), *Discovering Wine Country: Tuscany* (2006, Mitchell Beazley), *Château Monty* (Portico, 2008), *Monty Waldin's Best Biodynamic Wines* (Floris, 2013) and *Biodynamic Gardening* (Dorling Kindersley, 2015). Monty has contributed to *Hugh Johnson's Pocket Wine Guide* (Mitchell Beazley), Tom Stevenson's *Wine Report* (Dorling Kindersley), and writes the entries for biodynamics, organics and natural wine amongst others in *The Oxford Companion to Wine* (ed. Jancis Robinson MW, OBE). He has also appeared on BBC TV and Radio (The World Service, Radio 4's *Food Programme*), and has contributed to British newspapers (*Independent, Daily Mail,* London's *Evening Standard*), and websites (www.jancisrobinson.com), as well as to wine, travel and environmental publications including *Decanter, World of Fine Wine, Le Pan Media, Harpers Wine & Spirit Trade Review, The Ecologist, Star & Furrow* (the journal of the UK Biodynamic Agricultural Association) and *Biodynamics* (the journal of the USA Biodynamic Farming and Gardening Association). Monty is a regular speaker at conferences on green issues and has advised vineyards and wineries in both hemispheres who are looking to green up.

INTRODUCTION

Wines are generally defined in one of three ways: by their country or region of origin, by their colour (red, white, pink) or by their style (still, sparkling, fortified). Only recently have wines begun to be defined according to how they have been grown and made. There is now a clear if over-simplistic divide between modern conventional or 'chemical' winegrowing on the one hand and 'green' alternatives on the other. These include 'natural' wine, an increasingly popular term now in both hemispheres but which is unregulated and banned in, for example, Italy, and organic and biodynamic wine, both of which are independently regulated. Organic and biodynamic vines comprised around 6 per cent (and rising) of the global vineyard in 2015, up from less than 0.5 per cent in 1999. The bulk of the organic vineyards are found in Europe's wine powerhouses of France (the Midi, Provence, Alsace), Italy (Sicily, Calabria, Tuscany) and Spain (Penedes, La Mancha), as well as Germany (Rhine regions, Baden) and Austria. In the southern hemisphere, Chile seriously underperforms for a country claiming to be a wine paradise (perfect winegrowing climate, few pests, free Andean irrigation, no phylloxera), and Australia does not bother to collect national stats on organics. In contrast, New Zealand has set itself the most ambitious target of any wine-producing nation, namely to have 20 per cent of its (currently 33,300 hectare) vineyard certified by 2020.

Nevertheless, sceptics deride biodynamics as an extreme form of organics, with its quasi-religious overtones and possibly even voodoo practices, and at best 'merely organic agriculture with a peculiar twist'.[2] Although neither its reliance on essentially mundane ingredients – like the animal manure, wild plants and minerals incorporated in the nine preparations which are prerequisites for biodynamics – nor its adherence to lunar and other celestial rhythms appear to harm either vineyard soils or wine drinkers, biodynamics nonetheless lacks, for such sceptics, a proper scientific foundation.[3] For advocates like me, however, biodynamics offers effective, creative, enjoyable, stimulating and sustainable solutions to common problems experienced by contemporary winegrowers, such as reduced soil fertility, vines' diminishing resistance to pests and diseases, and grapes which, despite being increasingly complicated to ferment, risk producing ever more banal wines largely devoid of individuality and interest. If consumers are to be successfully encouraged to change their environmental habits then eco-warriors must play a part by altering their relentlessly downbeat message and offer instead a happy future to look forward to rather than a bleak one to avoid.[4] Biodynamic wine is one of those often everyday yet also conveniently rarefied products capable not just of encouraging that change, but of leading it.

FOUR WAYS TO FARM GRAPES

Traditional subsistence

From around 12,000 years ago hunter-gatherers began domesticating plants and animals. Traditional subsistence farms emerged. In terms of resources, says biodynamic farming consultant Andrew Lorand,[5] subsistence farms make above-average use of what is already available on the farm or grown there, so little needs to be bought in. In wine, for example, rather than purchasing wooden support posts for vines, fruit trees can be planted as supports instead. Their fruit also provides an extra crop. However, because the traditional subsistence system produces even fewer outputs than inputs, little if any of the farm's produce being sold or even bartered, ultimately this system is not economically sustainable. The closest modern-day equivalents of traditional subsistence vineyards – still a feature in some Latin American, Mediterranean and eastern European countries – form part of mixed smallholdings or farms in which wine is a minor activity. Any wine produced is for the family table rather than for sale.

Industrial

Modern industrial farming and winegrowing were spawned by the agricultural and industrial revolutions. In the industrial system any resources either already available on the farm or grown on it are poorly used. An example would be if prunings or grape residues left over from winemaking each year were burnt rather than recycled as compost, or were left to compost by neglect on the vineyard in ways that might actually increase pest and disease problems later on. Although the modern industrial system is capable of producing huge outputs – lots of bottles of wine – this is only at the expense of even greater inputs. In the vineyard such inputs might include soluble fertilizers to boost yields and man-made sprays to control weeds, pests and diseases. In the winery aids, additives and agents, like enzymes, acids, tannin, sugar, yeast and yeast food, can be purchased to compensate for grapes lacking colour, freshness, texture, ripeness and fermentability respectively. Most contemporary wines, whether sold in bottle or bulk, result from this system. The vineyard and winery become conduits through which myriad materials pass, few if any of which enhance the most precious and irreplaceable resource of all for a winegrower, namely the soil.

Organic

Organic winegrowers have promoted the organic system as producing superior, potentially more authentic-tasting wine because organics stipulates that no synthetic or man-made products may be used. Organic growers maintain that as their system is less likely to leave unnatural residues either in the soil or the wine it must therefore be considered sustainable. The numbers of organic vineyards worldwide have risen consistently since the mid-1990s, albeit initially from a very low base and encouraged initially more by state subsidies than consumer demand (consumer perception that 'all wine is organic' has proved hard to shift) or winegrower altruism. Recently rises have become especially significant in France (Alsace, Beaujolais, Languedoc, Roussillon, Provence, Loire, Rhône, Jura), Austria, Italy, Spain (especially Penedes) and New Zealand. Worries over soil erosion, spray residues in wine causing problems in export markets, reduced groundwater quality, pest and weed resistance to expensive conventional sprays, and potential lawsuits from angry parents (as in Bordeaux in 2015) worried that spray drift near schools is damaging the health of their children have contributed to the fear factor. On the positive side there is increased market demand for more 'terroir-driven' wines of higher quality both from fusty state alcohol monopolies and trendy 'natural' wine bars, and on the supply side the domino effect is now clear – the pioneering organic neighbour has morphed from being ridiculed as the lone mad wolf to being something of a break-the-mould pioneer by showing that you can be green in the vineyard whilst staying out of the red in the bank. Market demand has also spurred advances in lighter, quicker and better spray machinery, better sprays and better analysis of climate data to predict pest and disease attacks.

However, Lorand argues that even though the organic system produces bigger outputs than the traditional subsistence system, this is only at the expense of more inputs too. For example, whereas in the industrial system a vine pest can be eradicated with just one pass of a tractor spraying pesticide, the organic system may require three passes of the tractor spraying herb-based pest irritants like neem, so more tractor diesel is burnt. However, the organic system at least makes better use of on-farm resources than its industrial counterpart, meaning, for example, that rather than being sent to landfill, any grape residues left over from winemaking might be composted and returned to the soil, albeit perhaps with some purchased compostable material like animal manure or straw. Although the organic system is more sustainable than its industrial counterpart, the organic vineyard nevertheless remains a conduit through which materials pass in a way likely to create an imbalance between what the organic vineyard produces and what it consumes to remain economically viable.

Ecological

Balance is achieved only in the ecological system because this is the only system in which outputs are at least equal to or are greater than inputs. This is because any available resources are always efficiently used. One example would be harvesting rain to irrigate crops or provide water for livestock. These livestock would provide both food for farmers (meat, milk, cheese, yoghurt) and enough food via their manure for both the soil and the crops. This manure would be combined with other farm waste in compost in a way that promoted sufficient soil fertility in the short term, with long-term soil enhancement overall. Crop and soil health could be maintained via medicinal teas and liquid manures made from wild plants growing around the farm. Maintaining surrounding biodiversity would encourage beneficial insects to participate in self-sustaining pest control across the farm. All these measures combine to help crops stay healthy and well fed at very low environmental cost. The ecological system imitates nature, and has four major principles as identified by Lorand.

1. Appropriate production, which means common-sense farming. From a wine perspective, appropriate production means growing the right grape variety on the right soil in the right climate in the right way. This may seem blindingly obvious, but in Europe, for example, it took centuries for first the Greek and Roman colonizers of antiquity and then religious (Cistercian monks) or secular medieval vineyard owners (sixteenth-century Bordeaux *parlementaires*) to match the right grape to the right *terroir*. Riesling to the Mosel, Pinot Noir and Chardonnay to Burgundy, and Cabernet Sauvignon to Bordeaux's warm left-bank sandy gravels and Merlot to its cooler right-bank clays and limestones. Vines which adapted to their local surroundings grew more healthily, were easier to work and cheaper to farm. They also produced the best and most valued wines. This is appropriate production. An egregious example of inappropriate production might be the Murray Darling basin in Australia, a desert artificially transformed into a green vine oasis thanks to near-hydroponic fertigation. The results of this are huge yields of cheap, characterless grapes offset by salination and silting of the once mighty Murray Darling basin so catastrophic that the government is spending billions repairing the environmental damage whilst buying back the usufructuary water rights it once all but gave away to farmers. This unusually uncuddly version of winegrowing appears especially inappropriate whilst fresh water supplies are diminishing worldwide and global wine consumption consistently falls short of global wine overproduction. Australia is also, lest we forget, the world's driest continent.

2. Biodiversity,[6] which means avoiding the monocultural approach of putting all your eggs in one basket. Biodynamic growers take a holistic view of nature, working with it rather than against it to promote the kind of biodiversity which helps establish a natural equilibrium between the farm and its surroundings. Biodiversity helps maintain a broad enough gene pool to render carriers of illness less potent,[7] reducing the risk that any single pest or disease can destroy entirely the farmed crops. Biodiversity is seen as a hard goal to achieve by those growing perennial crops like vines or fruit trees because these are usually grown as monocultures, meaning that in any given field just wine grapes or apples and no other crops are grown, sometimes for decades on end. The rotation arable farmers practice – whereby fields

are ploughed and sown every few months with different arable crops like clovers which put food in the soil and grains which then take this food out again – is impossible. In a perennial vineyard monoculture the same vine grows in the same place for many years and therefore is forced to live in its own waste, its leafy detritus, increasing the potential risk of disease. To make vineyard mechanization and spraying easier usually only one single variety of wine grape is grown in any one field, because planting an early ripener like Merlot with a late ripening one like Cabernet Sauvignon would complicate pest and disease control, canopy management and picking decisions. Single grape monocultures are often enshrined in national wine laws in Europe, e.g. 100 per cent Chardonnay in Chablis, 100 per cent Sauvignon Blanc in Sancerre, 100 per cent Sangiovese Grosso or 'Brunello' in Montalcino and so on. The most extreme example of a monocultural vineyard results if just one single clone of the same variety – clone 114 of Pinot Noir, for example – is planted in a single plot. Vineyard monocultures can be described as 'environments without companionship'. Transforming them back into more diverse habitats is hard but worthwhile work which requires thought. Beneficial insect predators and parasites are attracted by pollen and nectar, so sowing companion plants like cover crops or leaving native plants (weeds) *in situ* to provide these things saves money on both weeding and pest control. The wrong type of biodiversity, however, can cause problems. Lucerne (*Medicago sativa*) or alfalfa can be sown as a cover crop because its deep tap root loosens soil compacted by tractors, releasing trapped boron (a trace element vines need to set fruit), but in California lucerne can act as a host plant to sharpshooters vectoring Pierce's disease, which is deadly to vines.

3. Soil fertility, which means having healthy foundations. Until recently soil used to be thought of essentially only in terms of its physical properties such as how well it drained or warmed up, or its chemical properties meaning how acid or alkaline it was (pH), or which minerals it contained. The vital role of soil biology was somewhat overlooked, ironically given Louis Pasteur's work in identifying the role different living organisms played in converting grape juice to wine and thence to vinegar.

In a living soil visible organisms like worms and beetles and microscopic ones like bacteria, fungi, nematodes and protozoa form part of a soil food web. Vines must be very active members of this soil food web to stay healthy. Vines capture the sun's energy via the chlorophyll in their leaves and use this to transform atmospheric carbon dioxide into food in the form of carbohydrates. Some of these carbohydrates end up in the grapes as sugar for wine, and some the vine releases via its roots. Soil micro-organisms like mycorrhizal fungi which live on the roots then feed off these carbohydrates. The mycorrhizal fungi return the favour by allowing vines to locate and capture soil nutrients. Plant roots can only assimilate soil nutrients in solution. By liquidizing or dissolving soil around the vine roots mycorrhizal fungi not only allow vines to feed, they also create spaces for vines to put down deeper, thicker roots. Thus vines and soil micro-organisms have a symbiotic relationship, each being reliant on the other. However, mycorrhizal fungi are killed by weedkillers and are made redundant by inorganic (soluble) fertilizers which, once dissolved by rainwater, can be taken up directly by plants. Vines which can no longer feed from the soil because of shallow, stunted or redundant root systems are more prone to pests, diseases and climatic disorders (drought, floods). They are also unlikely to be able to transmit that 'somewhereness' which drinkers of top Bordeaux, Burgundy and other *terroir*-driven wines expect to experience (see Chapter 6, on compost teas) and often increasingly if erroneously describe as 'minerality'.

4. Immunological response capacity, which means 'not getting ill'. If the right vine variety is being grown on a vineyard site to which it is suited and if the site/vine are maintained in a biodiverse way above ground with additionally a living soil below ground, then the vines should exhibit an immunological response capacity. In simpler terms this might be termed the 'apple a day' approach. Humans who rarely fall ill could be described as having self-sufficient immune systems by virtue of living in stimulating, unpolluted environments appropriate to their lifestyle (right *terroir*), eating a balanced diet (biodiverse, worm-rich soil), and exercising regularly (strong roots). Similarly, vines able to draw from the atmosphere via photosynthesis and from the soil via their roots exactly what they need to grow healthily exhibit self-sufficiency. If every vine in the vineyard were self-sufficient, then by definition the vineyard would be too; and if the vineyard were self-sufficient with minimal or no external inputs needed, but was still able to produce a net output of wine and possibly other crops as well and as part of its biodiverse orientation, then there would be a net gain both for the winegrower and the natural ecosystem. Growers would be able to produce

a high quality product – healthy grapes produce the best wine – whilst putting more back in than what had been taken out as grapes. This is what biodynamics sets out to achieve: healing the earth while producing healthy, vital and flavourful crops.

The 'don't get ill' approach can be summed up by Graeme Sait, a globally renowned expert in sustainable agriculture and horticulture. He says, 'Crop resilience is multi-faceted. There are no magic bullets. If you can work towards managing your minerals and microbes, eliciting an immune response, boosting photosynthesis and strengthening cell walls with calcium and silica you will reduce your requirements for increasingly expensive farm chemicals. More importantly the mechanisms governing resilience are the same things that govern yield and quality. An holistic approach like this can also boost production and profitability.'

It is my firm belief, based on practical experience working first in conventional, then in organic and finally in biodynamic vineyards and wineries on and off since 1984, that biodynamics remains the best tool with which to make *terroir*-driven wine of the highest quality while enhancing rather than depleting the vineyard it came from. The very first biodynamic vineyard I visited in 1993 in Bordeaux's Canon-Fronsac sub-region convinced me of this. For me the biodynamic tool remains as valid now as it was then. Not all biodynamic winegrowers use the tool as well as they might, not all biodynamic growers make great wine, and quite a few newer converts clearly see it as a potent marketing tool in our ecologically aware, environmentally challenging but still cynical times – but that's the fault of human beings, rather than of the biodynamic idea itself.

1

THE ORIGINS OF
BIODYNAMICS

Biodynamics dates from 1924 and is the oldest alternative agriculture movement. Biodynamics pre-dated the global organic agriculture movement whose founding organization, the UK's Soil Association, dates from 1946. In fact the very word 'organic' was derived from the biodynamic ideal that each farm or smallholding should always work towards becoming a self-sustaining *organism* in its own right.

The particular feature of biodynamics – and where biodynamics differs from organics and indeed all other forms of alternative agriculture – is the use of nine so-called 'biodynamic preparations'. These are made from cow manure, the mineral quartz (also called silica), and seven medicinal plants: yarrow, chamomile, stinging nettle, oak bark, dandelion, valerian and *Equisetum arvense* or common horsetail. These nine preparations are applied to the land or crops either by being first incorporated into a compost pile or by being diluted in water as liquid sprays.[1]

Biodynamic preparations are used in homeopathic quantities, meaning they can produce an effect in extremely diluted amounts, but they are not homeopathic treatments per se. Their purpose is to make the farm and farmer, its crops, animals and wild habitat, self-sufficient, self-sustaining and socially, economically and spiritually robust. These concepts may seem woolly in our world of smartphones and space exploration, but would have seemed less so to 1920s Europeans coping with the ravages of both the First World War and then its even deadlier successor, an influenza pandemic.

The methods used to make some of the preparations may seem strange initially but are neither high tech, expensive, costly to the environment nor potentially harmful. Anyone, from children to grandparents, can (and do) make these preparations. The biodynamic preparations are not patented so they can never realistically be made purely for profit, and they seem to get good results for farms and vineyards.

Sceptics, however, claim the biodynamic preparations produce no measurable changes to either farm health or crop quality; thus there is no 'biodynamic effect'. Such sceptics argue that biodynamic winegrowers owe the high quality of their wine either to having a top-quality vineyard *terroir* in the first place, or that vineyards which improved after 'going biodynamic' did so because the winegrower learnt to become extra attentive in the vineyard by following a biodynamic 'prevention rather than cure' mindset (e.g. better pruning, recalibrating spray machinery so sprays are more effective), and not because the vines were treated with biodynamic sprays or composts. Nevertheless, increasing numbers of winegrowers are using these preparations which are essential to biodynamic agriculture. Their regular use is the fundamental requirement of Demeter, the non-profit organization which has overseen and certified biodynamic agriculture worldwide since 1928 (see Chapter 8).

The biodynamic preparations were created by an Austrian called Rudolf Steiner (1861–1925) shortly before he died. His motivation was to remedy what he sensed was the arrested spiritual development of his contemporaries. Steiner believed the forces people needed to kickstart their spiritual development would come from digesting food imbued with these desirable and necessary forces, and that getting these forces into food required a new way of growing food: biodynamic agriculture. For this Steiner developed nine biodynamic preparations to moderate and regulate biological processes in nature. This is the 'bio' part of biodynamics. The 'dynamic' part comes by understanding the preparations' role in enhancing and strengthening forces that form or shape material substance, both on the farm and within both the farmer and the crops. These forces are referred to in biodynamics as 'etheric formative forces'. Like gravity, they are unseen but have a tangible effect on both soil and on crop plants as well as on the animals or humans who digest those plants. Steiner's nine biodynamic preparations can therefore be thought of as spiritual remedies for the human being which are administered indirectly through the healing process of the Earth.[2] Biodynamic farmers accept that there is no substance or matter without spirit, and equally no spirit without matter. So the point of growing biodynamic food and drink is not only to provide the substances (vitamins, carbohydrates, protein, fats, minerals) to nourish the human body but also to provide the forces needed to form and nourish the human spirit.

The spiritual side of biodynamics is the one most open to misinterpretation, misrepresentation and ridicule. One common misconception is that apart from encouraging you to start wearing sandals and paying less attention to personal hygiene, growing or eating biodynamic food will also turn you into a religious fruitcake. I discovered biodynamics in 1993 but had struggled to find many redeeming features in organized religion from the age of seven (1974) onwards. I am not a fan of sects. I do consider myself spiritual in the pantheistic sense of feeling my spirit lift palpably when I feel a connection with the natural world. This can happen when standing euphorically on the top of a mountain or, more mundanely, when looking at pigeons fluttering around under the eaves of the railway station my train is about to depart from.

In my experience winegrowers – be they biodynamic or conventional – who come across as fundamentalist proselytizers tend not to make the best wines, often because they are inflexible and unwilling to compromise. This may be fine when churning out widgets on a production line but is not adapted to a product like wine, dependent on the vagaries of nature. Fortunately, the fundamentalist proselytizers tend to be in the minority.

Most winegrowers newly adopting biodynamics start by seeing it as I did initially: as a sensible, doable, interesting, inexpensive tool to produce tastier grapes to nourish the human palate – and if they also provide the formative forces to nourish the human spirit, so be it. Biodynamic ultras argue that this purely 'substance rather than forces' way of looking at biodynamics means missing the real reason we should be biodynamic. I would argue that materialistic and only vaguely spiritual people like me – meaning exactly the kind of people Steiner developed his biodynamic preparations for – first have to understand and accept how the biodynamic tool works, and only then can we perhaps appreciate that our spiritual development may have lacked something to begin with after all.

STEINER'S 1924 *AGRICULTURE* COURSE

Steiner first described the biodynamic preparations publicly between 7 and 16 June 1924 during a course on agriculture consisting of eight lectures and four discussion sessions. This series of lectures was published in English (various translations) as *Spiritual Foundations for the Renewal of Agriculture* but is more usually referred to in biodynamic circles simply as the *Agriculture* course.

The *Agriculture* course was held at Count Carl von Keyserlingk's estate at Koberwitz near Breslau in what was then Silesia in the eastern part of Germany but is now Wroclaw in Poland. More than a hundred farmers, vets and others whose livelihoods depended on the land attended. They had asked Steiner to give the 1924 *Agriculture* course partly because their livestock was suffering ever more frequent outbreaks of foot and mouth disease, but also because they could see small, diversified family farms being swallowed up seemingly inexorably into much bigger, more mechanized and overtly monocultural ones. The area around the Koberwitz estate had been especially affected by this trend. The attendees were all members of the Anthroposophical Society, which Steiner had founded in 1912.

Anthroposophy

Anthroposophy or spiritual science is a view of life that includes both spirit and matter. It sees plants in a slightly unusual way, as having four organs to their development: the root, the leaf/shoot, the flower and finally the fruit, meaning the part which contains the plant's reproductive seed. These archetypal four organs also relate to the four elements (or ethers, see p.150): the roots to the earth, the shoots and leaves to water, the flowers to air/light, and the fruit/seed to fire or heat.

Thinking of individual vines or farm crops as being made up of four organs is the first step in understanding entire vineyards or farms as being individualities or organisms in their own right. When anthroposophical farmers picture the agricultural individuality or farm organism they do so as though it were a person standing with his/her head in the soil.[3] Weleda, a German-based pharmaceutical company (www.weleda.com), pioneered the development of anthroposophical medicines from 1921 by following Rudolf Steiner's advice, selecting specific plant 'organs' to treat specific human conditions. Plant roots, which compare to the human head, provide remedies for problems in the nervous (senses) system. Leaves and stems provide remedies to treat the human rhythmic organism (heart, lungs, circulation), while the fruit or flowers serve the sexual and metabolic (digestive) system. Healthy farm organisms should be resilient, self-sufficient and produce not just healthy crops but farmers of healthy mind and body too – as though the farmer and his animals are running around in the belly of the farm.[4]

Steiner's path to anthroposophy

This anthroposophical way of thinking about plant organs and the farm organism was developed by Rudolf Steiner. He was born in 1861 in Kraljevec, then part of the Austro-Hungarian empire and now part of Croatia. His father Johann was a station master on the newly constructed Vienna–Trieste railway; his mother Franziska was maid to the local count. During his early childhood Steiner developed a strong connection to nature via local smallholders, peasants whose feudal existence had remained essentially unchanged for centuries. Steiner's connection with the natural, physical world around him was matched by the connection he sensed with an unseen, spiritual world that lay behind it, a world he felt a need to explain or codify in some way before it slipped inexorably away forever from both him and his contemporaries. Steiner believed he had a clairvoyant ability to connect with the natural world but was well aware that modern methods of communications, like the newly installed telegraph his father used daily for his job on the railway, would soon render anyone like him, who claimed to sense an inner perception of the non-physical world, ridiculous.

By his eighteenth birthday Steiner had moved with his family closer to Vienna, where he studied science at the city's Technical Institute, then considered one of the world's foremost scientific universities. Steiner's humanistic, spiritual side may have risked being extinguished in this increasingly science-oriented world but for a chance meeting on the train taking him to school with Felix Kogutski, a herb-gatherer who sold medicinal plants both to the city's pharmacies and the botanical department at the medical school. Steiner felt the course of his life changing because he believed Kogutski represented 'an instinctive clairvoyance of an earlier era'. Kogutski provided Steiner with the opportunity to talk with a like-minded person about the spiritual aspect of reality.

Steiner switched from science to arts, leaving Vienna for Rostock in Germany where he studied literature and philosophy, entitling his dissertation 'Truth and Knowledge'. After publishing an introductory book called *The Theory of Knowledge Implicit in Goethe's World-Conception* in 1886, Steiner was invited by the Grand Duchess Sophie of Saxony to edit the scientific writings of Johann Wolfgang von Goethe (1749–1832), Germany's most famous writer and poet. Goethe's pioneering work in phenomenology and the organic sciences provided Steiner with what he had yearned for, the bridge between the seen physical world and the unseen spiritual world.

Goethe's phenomenological approach held that through regular observation of plants, animals or other living organisms in all stages of their growth, inner and outer pictures of their processes of movement and their changes of form can be developed, and that insight into natural laws and processes could be gleaned.

Goethe's approach asks us to think about how plant growth is affected by intangible but identifiable 'nitrogen processes' rather than merely by how much nitrogen a plant has been fed with, which is the approach of physical chemistry. While working on Goethe's archives in Weimar, Germany from 1890 to 1897, Steiner published his own seminal work called *The Philosophy of Freedom* (1894). Its anti-materialist thrust, that humans become spiritually free only through the conscious activity of thinking, was the basis of Steiner's own theory of spiritual science or 'anthroposophy', from the Greek *anthropos* or 'wisdom of man'.

As well as biodynamic agriculture, anthroposophy embraces diverse fields. Waldorf education is one example, Steiner having been asked in 1919 to formulate his educational theories by Emile Molt, a representative for workers in the Waldorf-Astoria cigarette factory in Stuttgart. An offshoot of Waldorf education is the Camphill Movement of schools and villages for children with severe learning disabilities, founded in 1939 by Dr Karl König. With his wife Marie von Sivers, Steiner developed eurythmics, a human art form with therapeutic uses sometimes referred to as 'visible speech'. In the field of medicine Steiner collaborated with pharmacists and physicians in creating Weleda (mentioned above), whose plant-based medicines paid homage to

Kogutski, the herb-gatherer on the Vienna train. Steiner also conceived an organic form of architecture when creating his school for spiritual science in 1913, the Goetheanum in Switzerland, in honour of Goethe (see Appendix II).

Von Liebig's 'Law of the Minimum'

What Steiner had understood from Goethe and his own intuition was the flaw he saw in modern scientific methods: science was limited when it came to helping man understand nature and, by implication, farming because it overlooked the spiritual aspect of reality. The most obvious examples of concern to Steiner were the inorganic soluble fertilizers popularized by Baron Justus von Liebig (1803–73), the German chemist regarded as the father of so-called 'chemical' farming. It was von Liebig who put forth the 'Law of the Minimum', the idea that whichever essential inorganic plant nutrient was least available to a plant would dictate or limit its growth potential.

The consequence of von Liebig's idea was the development of soluble nitrogen, phosphorus and potash (or 'NPK') fertilizers. Via the Haber-Bosch process, named after its German inventors, chemists Fritz Haber and Carl Bosch, nitrogen in the atmosphere could be converted under high pressure and high temperature to ammonia (nitrogen comprises nearly 80 per cent of the atmosphere, oxygen making up most of the rest). The process was initially used to make explosives during the First World War to compensate for the loss of saltpetre whose supply from British-owned mines in Chile had been blocked by the Allies.

Steiner argued that using these artificial fertilizers weakened crops by forcing plants to expend some of their own life energy or vitality in raising these lifeless chemical substances from the inert mineral state to an alert one capable of carrying life force.

However, artificial fertilizers – although expensive – conveniently boosted farm yields at a time of rapidly rising and shifting post-industrial populations, as small rural communities were superseded by large cities. Artificial fertilizers were there to stay despite unwelcome side effects, like producing bigger, more persistent weeds. They were also detrimental to the balance of key soil micro-organisms, making it harder for plants to feed as normal via the fungi on their roots and for the soil to maintain a viable, healthy structure. This, coupled with the increased use of steel ploughshares pulled quickly by machines rather than more slowly by animals, exacerbated soil erosion – the 1930s dust bowl in the American Midwest becoming the prime example. As crops weakened, so did the animals and humans fed on those crops, as Steiner had predicted. Steiner said water-soluble or mineral fertilizers as he called them would encourage excess fungal activity in the topsoil, and that this would result in the migration of the excess spores onto crops which would then became more prone to disease.

Even before he died von Liebig had begun to rue his attempts at playing chemical God to the soil and the negative effect his work had had on farming. Were he alive today, von Liebig would have realized how plants are able to assimilate only about 10 to 15 per cent of the nitrogen provided by soluble fertilizers, and that the excess free-nitrogen leftovers contribute significantly to global warming (via nitrous oxide, a more potent greenhouse gas than carbon dioxide, and one which hangs around for longer), ozone-layer depletion and acid rain. Phosphorous, potassium and nitrogen run-off from fertilizers also stimulate algae to breed frenetically, causing algal blooms which create so-called 'dead zones' in rivers, lakes and the sea. The dead zone in the Gulf of Mexico, caused by run-off from Midwestern farms, is now larger than the state of Connecticut.

Scientific substance versus spiritual force

The farmers of the Anthroposophical Society who came to hear Steiner's *Agriculture* course at Koberwitz in June 1924 knew that the anthroposophical plant-based medicines Steiner had developed with Weleda provided ways of using wild plants to heal humans. What they wanted to know was whether Steiner could describe a way of reversing the declining health of their livestock, crops and soil too. Steiner made it clear from the outset that rather than provide a Liebig-like way of farming by numbers his aim was to provide a new way of thinking about farming, food and nutrition. His message was that we are what we eat, certainly, but we must not think of food by simply calculating how much we consume in terms of grams or calories, especially since most of what we eat we then excrete anyway. We must think about whether our food contains the forces we need to nourish our spirits, and maintain both our vitality and that of the planet we live on.

Steiner's precise words were, 'The most important thing [is] to make the benefits of our agricultural [biodynamic] preparations available to the largest possible areas over the entire earth, so that the earth may be healed and the nutritive quality of its produce improved in every respect ... This is a problem of nutrition. Nutrition as it is to-day does not supply the strength necessary for manifesting the spirit in physical life. A bridge can no longer be built from thinking to will and action. Food plants no longer contain the forces people need for this.'[5]

The implication was that if we dumb down our farming we dumb down our food, and seeing as we are what we eat we'll risk entering a vicious circle of thinking that there's actually really nothing wrong with dumb, even dangerous, farming – dangerous in the sense it threatens our very existence.

A pertinent example of this very risk is the 'terminator' or 'suicide' seed technology developed in the United States whereby second-generation seeds are either sterile or need to be coated with a commercially patented compound to become capable of reproduction. Farmers must therefore either pay for new seeds each season or pay to activate saved seeds. Farmers who find patented genes have migrated into their own seeds are at risk too of being threatened by the patent holder for theft of patented property. This is like being sued for the theft of paraffin and matches by the stranger who just used them randomly to set fire to your house. So rather than face up to a multinational, farmers find it easier to switch to patented seeds, meaning the seed company has a client for life.

In contrast, biodynamic agriculture aims to make farmers as independent, as self-reliant and as self-sufficient as possible. They are encouraged to save seeds from crops of the same type which have been open-pollinated naturally, by letting plants flower and then become fertilized by wind-borne pollen. In this way plants exchange characteristics from generation to generation but breed true to type, meaning the saved seed will always closely resemble the parent plants and pass on their characteristics whilst maintaining genetic diversity. Thus each new generation of fertile seeds can be sown the following season. Crops from open-pollinated seeds are both sustainable and genetically unique. This contrasts with both genetically modified seeds and F1 hybridized seeds like the ones you find in the supermarket or garden centre. The enforced inbreeding means these seeds will eventually produce unviable crops if left to set seed in future generations, meaning new seeds must be bought.

The seed as an image of the whole universe

Seed saving (by open pollination) has been part of farming since people ceased hunter-gathering at the end of the Neolithic, roughly 10,000 years ago. Rudolf Steiner said that 'in the seed we have an image of the whole universe'.[6] As today's seed is tomorrow's food it stands to reason that terminator seed technology implicitly promises humanity a barren future, a world in which nature is selectively patented by the few and ransomed back to the many at our collective expense.

The biodynamic alternative is a free, safe, unpatented and therefore universally available technology reliant on cows, some wild plants and a few handfuls of the world's most abundant mineral. By working with, rather than against, nature to resolve problems, biodynamic farmers see themselves as but one tiny part of a much bigger cycle of life, both guiding nature and being guided by it.

Ecology is rapidly shifting from being a good idea to being a matter of life and death as we face the combined challenges of climate change and the depletion of both biodiversity and basic natural resources like fresh water. Our fellow humans can increasingly be divided up between those who eat way too much and eat the wrong stuff – it takes eight kilos of grain to make a kilo of meat – and those many more who get to eat way too little. Writer Michael Pollan's much quoted dictum that we should 'eat food, not too much, mostly plants', is the perfect starting point for improving the environmental and physical health of our surroundings. The United Nations reports that there is no more potentially farmable land left to develop and that the dramatic yield increases of the type witnessed since the 'green revolution' of the 1960s onwards are essentially over. This means we'll have to make do with what we have already, but without further degrading, eroding or polluting the land. We'll have to become better and more self-sufficient stewards of what land we already farm.

The biodynamic principle of low-input, self-sufficient agriculture revitalizing both for ourselves and our natural surroundings seems useful, necessary and perhaps most conveniently of all, easily achievable.

2

THE BIODYNAMIC PREPARATIONS 500–508

Rudolf Steiner created the nine biodynamic preparations from natural substances – cow manure, quartz and seven medicinal plants. Making the preparations involves sheathing six of these substances in specifically chosen animal organs.[1] We may find this discombobulating, but Steiner's contemporaries were quite used to butchering farm animals and game at home.

Understanding each individual preparation becomes easier if one thinks first of the original substance it is made from (mineral or plant), followed by whether the substance (sheathed or not) undergoes its transformation by being buried in the earth or hung in the air or soaked in water of the right quality (see Chapter 4), then the time of year it is made, and finally the period(s) of the year through which it is left to transform. Biodynamicists often follow lunar and other celestial cycles when making their preparations (see Chapter 7).

In the preparation profiles which follow, the three biodynamic field sprays are described first, followed by the six compost preparations. The stirring or dynamizing process which four of the nine preparations undergo is explained in Chapter 4. Beforehand it is worth noting how code or lot numbers can also be used when referring to the preparations.

THE PREPARATION NUMBERS 500–508

Biodynamicists refer to the preparations either by name or by a three-digit number, the latter simplifying things when having to communicate in a foreign language. This book uses both the name and number together for clarity. One theory suggests that because during the Third Reich (1933–45) the biodynamic preparations, as well as the German biodynamic farmers' group (*Demeter Bund*) and its research body (*Forschungsring*) which Steiner had suggested be created, were forbidden in Germany, those farmers and researchers who were working with biodynamics decided to code the nine preparations – with 500 for horn manure, 501 for horn silica, and so on – to avoid discovery. In fact, the numbers assigned to the biodynamic preparations are lot numbers dating from when Weleda (see Chapter 1) began cataloguing its plant-based anthroposophic medicines for humans. Weleda listed the biodynamic preparations as anthroposophic remedies, albeit for the soil and crops rather than directly for humans.

THREE FIELD SPRAYS – 500, 501 AND 508

Table 1: A summary of field sprays

Preparation	Horn manure	Horn silica	Common horsetail as a fresh tea	Common horsetail as a fermented liquid manure
Number	500	501	508	508
Made from	Cow manure	Silica, quartz	*Equisetum arvense, Casuarina* or similar	*Equisetum arvense, Casuarina* or similar
Animal sheath	Cow horn	Cow horn	None	None
When buried	Autumn to spring	Spring to autumn	N/A	N/A
Why used	Soil health	Crop taste	Vine health	Soil health
Stirring time in water (minutes)	60	60	20-60	20-60
Volume of prep per hectare per year[2]	300g	5g	100g	100g

For a long, healthy life, a vine's requirements are similar to those of a house: deep foundations, a roof that withstands the elements and strong supporting walls in between. The three biodynamic field sprays – horn manure 500, horn silica 501 and common horsetail 508 – assist in providing those metaphorical needs. The horn manure 500 soil spray allows strong foundations to form via a strong and complex root system. The horn silica 501 atmosphere spray encourages strong and vertically erect fruiting wood. Common horsetail 508 mediates between the two, being rather like a damp-proof course when sprayed either around the base of the house (as a liquid manure soil spray) or like an air brick to keep the paintwork from going mouldy (as a fresh tea sprayed directly on the vines).

Horn manure 500

Horn manure 500, or the horn dung or cow dung preparation as some call it,[3] is the one preparation that those with only a passing interest in or awareness of biodynamics seem to have heard of. Horn manure may even be considered the nursery-slope biodynamic preparation, because having mastered how to make it – bury a cow horn filled with cow manure for six months over winter then dig it up and extract the manure from the horn – and understood why it is used and how it is used in the way that it is, the rest of biodynamic theory and practice no longer seems quite so impenetrable. To others horn manure is an enigma. To my knowledge, science has yet to explain why cow manure stuffed into cow horns and interred from autumn to spring transforms into dark, humus-rich material which is pH neutral[4] and endowed with especially high levels of microbial life.[5] Steiner called this preparation the 'spiritual manure' and implied that science would only discover why horn manure exists not simply by analysing the chemistry or biology of the cow, her manure, her horns, the soil around the pit or the pit itself but by looking at how horn manure in particular and biodynamics in general was 'always conceived out of the totality', meaning a combination of both physical substances and what he called etheric formative or 'life' forces.[6] Hence giving the land horn manure 500 was akin to giving it a concentrated manuring force (see scientific substance versus spiritual force, p.8).

Of all the animal manures, that from the cow appears unrivalled in the powerful yet efficient way it stimulates soil fertility, plant growth and thus life (for some reasons why, see Chapter 3). From a biodynamic perspective, manure is not simply vegetable matter that has broken down whilst passing through an animal's digestive system but vegetable matter which has also been impregnated with an animal's metabolic forces.[7] Digestion releases energy as both intangible forces and tangible substances to nourish both the spiritual and physical body. Animals have no need, however, of the forces which human beings must develop in order to attain self-consciousness – as sentient beings we know we are going to die, but cows appear to be unaware their lives are finite – and they therefore release those unused forces.[8] In the case of the stag, for example, these metabolic forces are released via its antlers (see yarrow 502, below). In the cow, however, these forces are withheld by the retaining effect of the cow's horns and hooves.[9] 'If you could crawl around inside the living body of a cow … right inside the cow's belly, you would be able to smell how living astrality streams inward from the horns. And with the hoofs, it is similar,' said Steiner.[10]

Steiner said that by burying the manure in the horn the life (etheric) and soul (astral) forces already present in the manure can be further enlivened by the presence of the surrounding earth.[11] Doing this in winter is quite deliberate. Winter is when the earth recharges itself on a substance level – think about all the potential for growth, visible only in summer, stored underground during winter in roots or seeds. Winter is also the moment when the earth recharges itself on a forces level by

Female cow horns are a key tool in biodynamics

breathing in life-giving formative forces from the cosmic environment, which Steiner said play a formative role in shaping all living things. Silica crystals form in worm galleries, showing the earth's desire to become crystalline and how it is most inwardly alive in winter.[12] This crystallizing process enables these life-giving formative forces streaming in from the celestial sphere to radiate into the horn manure 500. The aim with horn manure 500 is to create something capable of impacting on the earthy, mineral component of the soil rather than on the soil's watery part which is the only part that water-soluble mineral or 'chemical' fertilizers are capable of reaching. Thus horn manure 500 is not fertilizer per se but provides a highly concentrated, life-giving, fertilizing force. It contains unusually high levels of enzymes, organisms which act like a living oil for the biological wheels of life. When you think of manure horns lying underground you can imagine them as being like giant worm castings whose role is soil rejuvenation.[13]

Horn manure 500 and the lime–silica polarity

Horn manure 500 is one of two biodynamic preparations made by using a female cow horn as a sheath, the other being horn silica 501 (discussed below). Both horn preparations are intended to enhance the ability of soil and crops to receive and actualize planetary forces in particular as well as cosmic forces in general. The two horn preparations can be seen as equals and opposites, especially since Steiner said that the oxides of calcium and silicon (lime and silica) represented the two opposite poles of life chemistry.

Horn manure 500 supports what Steiner called the earthy or lime/calcium principle, while horn silica 501 supports the opposite cosmic or silica principle. Horn manure 500 works with the etheric or life-force energy of the earth itself, and comes to expression in the growth of the plant, working on the plant 'directly' and in a 'building-up' way below ground, mobilizing the roots and stimulating humus formation, allowing the forces of the nearer planets (Moon, Mercury and Venus) to help plants grow and reproduce. Balance comes with horn silica 501, which ensures that whatever is produced in the way of crops is ripe, tasty and healthful.

Lovel calls horn manure 500 the quintessential humus,[14] the basis for intelligence and self-awareness, a central nervous system for the farm organism; the horn silica 501 then provides the sensory organs so that this self-awareness can be expressed and the farm manifests its own farm individuality. In wine-speak this means that each biodynamic vine, in fact each biodynamic grape, should be capable of expressing its own micro-*terroir* via its wine. This contrasts with our normal approach of seeing *terroir* expression as collective, a 'whole vineyard rather than individual grape or vine' thing. François Bouchet told me horn manure 500 'is what allows to vines create their own sense of self and self-expression, their *moi* [me-ness] if you like, by expressing themselves vertically downwards underground via their roots. It is the very basis of the concept of biodynamic *terroir*.'

Peter Proctor says that by improving both structure and humus levels in soil the horn manure 500 helps increase the soil's water-holding capacity by making it more permeable and humic, and balancing soil pH by lowering it in alkaline soils and raising it in acid soils. The soil, whether dominated by clay or peat, will in time take on the same crumb structure, he says, so wherever one is in the world one should be able to tell if horn manure 500 is being used or not. Horn manure 500 helps vine roots grow longer, deeper, thicker and more spread out, breaking up hard pans, and allowing improved nutrient uptake and resistance to climatic stress, especially drought.[15] Plant sap circulates more regularly, aiding primary shoot development. Horn manure 500 also attracts and stimulates beneficial soil micro-organisms like earthworms, azotobacter, bacteria and fungi, for example. Rhizobacter, bacteria which attach themselves to soil roots, become more active in soils where horn manure is being used, bringing increased nodulation in clovers, beans, peas and other legumes, and stimulating the germination of seeds like cover crops. Horn manure 500 helps regulate levels of lime and nitrogen in the soil, halting winter decomposition when used in spring, stimulating the release of trace elements just when the vines need them most. Horn manure 500 seems to work quickest in soils which are nearly permanently warm, says Proctor.

All of the above changes that horn manure 500 engenders in soil can be linked to the increased levels of oxygen it promotes. As Richard Thornton Smith points out,[16] oxygen is a key part of alumino-silicate minerals, better known as clays. On a physical substances level, oxygen plays a role in maintaining soil friability by aiding drainage, worm activity and plant root development. On an intangible forces level, oxygen allows formative forces which stream in from the celestial sphere and which are breathed in daily (afternoon) and seasonally (autumn–winter) to shape and form plants. Oxygen will account for at least 90 per cent of the volume of soil solids, clays and humus due to the large amounts of internal space in their layered structures. Formative forces drawn into the structures of humus and clay bathe nutrients which, when the earth breathes out daily (morning) and seasonally (spring–summer) allows these substances to be carried into crop plants and thus into our food. 'By having a distinct biography these nutrient ions will have a different *quality* [author's italics] from those deriving from NPK fertilizer,' Thorton Smith concludes.

Making horn manure 500

When choosing the horns, Steiner suggested it was 'best to use horns from your own locality. There is a very strong kinship between the forces present in the cow horns of a given area and the other forces at work in that area; the forces of the foreign [non-local] horns may conflict with the local forces of the Earth.'[17]

The horns should come from female cows which have had a number of calves, as seen by calving rings. These are formed when active lactation starts after calving and a constriction develops at the base of the mother cow's horn. This constriction continues to grow in length rather than in width, and calving rings develop.[18]

Cow horns which still contain their bony inner core can be hung in sunny shade for a week until the core comes away before being discarded. Cow horns are relatively heavy, thick walled and slender in form, and are hollow only about two thirds of the way to their tips. Brinton concluded that the bigger the weight:volume ratio of the horn used in making horn manure 500 and horn silica 501, the better the quality of the preparation.[19] Perhaps this is why Steiner said that bull horns were not suitable. Unlike cow horns, bull horns show no spirals, are conical in shape, thinner, ringless, larger in diameter and hollow all the way to the tips. Cow horns, in contrast, show some degree of spiralling right to the tip and their vortical shape suggests an ideal geometry for focusing the living energies of the earth on manure placed inside. For the significance of vortices and spirals, see Chapter 4.

Stuffing cow manure for horn manure 500 using a wooden dowel to prevent air gaps

For the manure, collect fresh cow pats in autumn directly from the pasture before the animals are moved inside for winter.[20] Ideally the cow pats will be firm, well-shaped (having a spiral form), and without straw, and from animals fed sufficient roughage by grazing on pasture or on a clover/grass mixture supplemented with hay and straw. Silage or beet leaf feeding is not recommended. The manure should come from your own land. Biodynamic cows never have their horns removed while alive and eat only a vegetarian diet.[21] The manure can be put into the horns fresh, on the day it is collected, or may be kept for up to two days (frost free) before being used. Manure can be stuffed into the horns by hand, with the help of a wooden spatula or dowel. Banging the horns every so often during filling helps the manure to reach right to the end of the available space.

The filled horns should be buried in a pit up to 75 centimetres deep. As horn manure 500 will take on some of the characteristics of the earth in which it is buried, the soil must be neither too clayey or too sandy.[22] The middle of a rich pasture treated with the biodynamic preparations for some time in which cows graze is ideal. The same pit can be used over and over again.

Freshly excavated horn manure 500 with the preparation pit behind

Bury the horns with the wide opening lower than the tip, to prevent rainwater seeping in. Pack the soil around each horn, burying them in layers if need be, so that every horn is surrounded by earth. There is no limit to how many horns can be buried in a single pit, as long as all the horns are covered by 30 to 45 centimetres of soil. Each horn should be fully in contact with the surrounding earth so the horns should not touch each other.

The filled horns are usually buried at autumn equinox and dug up at spring equinox six months later. The horns thus remain buried during the entire winter, the season when the earth's forces are inwardly directed (centripetal), as Steiner indicated. Ideally this 'earth remedy' is buried in the afternoon and under a descending moon, and for those following the sidereal moon as well when this stands in a root/earth constellation: Virgin in the northern hemisphere and either Goat or Bull in the southern hemisphere.

When the horns are dug up six months later their contents should be dark brown to brownish black in colour, supple, colloidal and elastic in texture,[23] and smelling pleasantly of humus. If the material inside is still wet, green and smells of manure, put the horns back in the ground for a few more weeks. If the colour is still greenish – due perhaps to water seeping into the filled horn – but the smell is right the most likely problem is a lack of air. The manure from such a horn will usually, if emptied and exposed to air for as little as twenty-four hours, acquire a more acceptable colour.[24] Horn manure 500 should always smell the same wherever in the world it is made, says Proctor, if one remembers that it is the bacteria, the life of the preparation, that give the smell.

Scrape away the earth from the outside of the horn before gently knocking the contents out to avoid inadvertently mixing the two. A wooden spatula can help dislodge the horn's contents. Steiner said horns for horn manure 500 'lose their forces after having been used three or four times'.[25] Between fillings, store them in the cow barn for the six months when they are not in use, ideally in a string sack which allows air to get to them. Chipped, damaged or rotten horns should be disposed of. As used horns are extremely difficult to break up they are usually buried beneath newly planted trees or shrubs, providing them with a long-term source of nitrogen.

Storing horn manure 500

Horn manure 500 is normally kept in glazed earthenware jars placed in a wooden box lined with peat and stored in a dark, cool, frost-free place. The lid of the jar should be loose fitting as this preparation needs to breathe.

Horn manure 500 ready for use – this is about enough to spray a single hectare of vines once

Fresh horn manure 500 should be made every year, although well-stored horn manure 500 which maintains a level of moisture will last up to three years. Small amounts left over from the previous year may be mixed in with fresh material.

Using horn manure 500

A useful rule of thumb is that the contents of one horn are enough for one hectare of vines per year, otherwise a figure of 30 to120 grams per spray is normal. Demeter's biodynamic standards suggest a total of 300 grams per hectare per year.[26]

The preparation is diluted in 25 to 120 litres of water, the exact volume depending on how much water is available, what kind of spray apparatus is being used to diffuse it and how much land there is to cover. Lower volumes of spray per hectare are usual on larger estates to save time and make handling easier.

The water is dynamized for one hour (see Chapter 4). Pierre Masson suggests stirring is best done in open air and light, in an area with good acoustics.[27] Crumbling the required dose of horn manure 500 in a small bucket of water before pouring it into the stirring vessel ensures it is thoroughly mixed, and avoids lumps of this preparation falling to the bottom where they will be neither stirred nor mixed.

Steiner suggested using warm water when stirring horn manure 500, warm meaning no warmer than the temperature of human blood

*Spray horn manure 500 on the soil in rain-like droplets
rather than as a fine spray*

(around 37ºC). The water should be warmed gently to the correct temperature by heating, rather than by adding boiled water to colder water, because water hotter than 37ºC loses its ability to assimilate and transmit life forces.[28] However, those who prefer cold water argue that the warmer the water becomes, the less able it is to absorb the oxygen which carries the life forces present in horn manure 500 onto the farm.[29]

Passing the dynamized horn manure 500 through a tea strainer on its way to the spray tank reduces the risk that any lumps or undigested grass seeds will enter spray apparatus where they might block the nozzles. Time unblocking nozzles is problematic, because once stirred the dynamized spray should be applied within four hours at the absolute maximum. It should be dripped onto the soil in the form of large droplets rather than as a fine spray which is the case with horn silica 501. Horn manure 500 can be mixed with some other soil sprays, notably stinging nettle liquid manure. When spraying by tractor, Pierre Masson suggests avoiding piston pumps, whose jerky return movements disturb the dynamized liquid, in favour of an electric diaphragm pump connected to the tractor battery. Attaching a speed regulator and manometer controls pressure and flow, he says.[30]

The prevailing weather and the moisture level of the soil are the two major practical factors determining when horn manure 500 is sprayed, and a far more relevant guide as to when to spray it than the lunar calendar. Aim to spray it on an overcast afternoon with little or no wind, and on soil which is moist but neither wet, frozen or likely to be heavily rained on immediately afterwards, and which has recently been or is about to be ploughed. In hot climates it can be sprayed in late evening, when the soil is warm but when the air is not so hot as to cause the preparation to evaporate before it reaches the ground.

As well as breathing in and out seasonally – in autumn and spring respectively – the earth breathes in and out on a daily basis too, exhaling in the morning and inhaling again in the evening.[31] This is a breathing in of forces rather than of air. The physical manifestation of this is the falling dew. Poppen says that the forces the manure acquired in the horn are transferred to the dew (if present) via the water it was stirred in.[32] Biodynamic growers report that there is often a marked increase in the dew the morning after an application of horn manure 500 anyway.

As the beneficial effect of horn manure 500 is enhanced by spraying it during the 'autumn' of the day, it is common practice to spray it in the autumn of the year too, as soon as possible after harvest, as both the seasonal and daily movement of the sun's arc in the sky becomes ever downward as the earth inhales. For this reason spraying when the moon is descending, and therefore in its autumn–winter phase, makes added sense. Growers will also try to spray horn manure 500 when the sidereal moon stands either in the root/earth constellations (Virgin in the northern hemisphere, and Bull or Goat in the southern hemisphere) or in the fruit–seed/warmth constellations (Ram in the southern hemisphere, Lion in the northern hemisphere, or Archer in either hemisphere). Peter Proctor recommends spraying horn manure 500 during a descending moon phase one to two days before moon opposition Saturn (see Chapter 7, moon opposition Saturn). Andrew Lorand sprays horn manure 500 during the first three or four months of the growing season,[33] from spring equinox to summer solstice, targeting the fourteen-day periods before the full moon when nature forces favour reproduction and growth which this preparation enhances. Its use is therefore essential when new vineyards are being prepared and planted. Lorand's timing does allow wider windows of lunar opportunity for spraying compared to Maria Thun's indications to follow the sidereal cycle, albeit with a greater risk of spraying horn manure 500 in hot midsummer conditions when both the air and soil will be dry (see p.33 for Lorand's views on when horn silica 501 should be sprayed).

Spraying horn manure 500 at least once is the bare minimum for a Demeter-certified biodynamic farm. Two or three annual applications are usual, one or two in autumn and a final one in late winter or early spring, before the earth begins to exhale once the sun's arc begins to widen and horn silica 501 becomes the horn spray of choice. Horn manure 500 can be sprayed intensively in the first few years of conversion to biodynamics.[34]

Rudolf Steiner was clear that spraying horn manure 500 was not to be seen as a substitute for spreading compost but as a means of enhancing the effect compost has on the land.[35]

Horn silica 501

Horn silica 501, or horn quartz (horn silicum) as it is sometimes known,[36] is the *yin* to horn manure's *yang*. While horn manure 500 influences the lower part of the vine and its roots, horn silica 501 influences the upper part of the vine, namely its shoots, leaves and the wine grapes. Whilst horn manure 500 is sent to work when the afternoon sun is approaching the yardarm and sprayed in thick drops on the dark and heavily tangible earth to pull vine roots downwards, horn silica 501 is mistily wafted into the bright intangible atmosphere to yank the vine shoots upwards as the morning cockerels are clearing their collective throats. Steiner said that horn silica 501 complements and supports the influence coming from the soil as a result of the horn manure 500.[37] Horn silica thus represents a concentration of the forces within sunlight.[38]

Von Wistinghausen *et al.* describe silica (quartz, SiO_2), 'the most beautiful form of which is rock crystal, [as being] the main constituent of the earth's crust (47%). In its pure form it is wholly translucent, hard, and water-insoluble. In spite of this silica is taken up into plants (grasses, horsetails) and the bodies of animals and humans (skin, eyes, nerves). It structures the soil (sand grains) and is found in aluminium silicate and in colloidal form in clay. Plants are able to take it up in colloid form. The atmosphere also contains finely dispersed silica. The fact that silica is found in sense organs, above all in the skin and in the eyes, makes us aware of its relationship to light. Industrial uses of quartz are in glass, optical instruments and for information technology. Very finely ground quartz has a large light-reflecting surface area. This is put into cow horns, which are then buried in the soil to expose the silica to the light and warmth of summer. When stirred in water for one hour and sprayed on to plants, this preparation conveys light qualities that have been transformed by the summer processes in the soil. This light energy promotes and organizes plant metabolism.'[39]

Horn manure 500 mobilizes matter in the plant which horn silica 501 forms and sculpts,[40] working on the internal structure of plants, favouring their uprightness or verticality (more visibly erect vine shoots) and strengthening the outer cell walls (epidermis) of vine leaves and grapes.[41] If horn manure 500 drives into wine a sense of place or *terroir*, the horn silica 501 ensures this sense of place tastes ripe.

Horn silica 501 and the lime-silica polarity

When asked in the discussion after the fourth lecture of his *Agriculture* course about the importance of silica, Steiner replied that, 'it is through silica that the actual cosmic factor is absorbed by the Earth and becomes effective'.[42] He also said that 'everything active in silica-like substances contains forces that do not originate with the Earth, but rather with the so-called distant planets – Mars, Jupiter and Saturn. These planets are working in the siliceous substances.'

Silica's relationship with these higher or so-called 'warmth' planets is said to give it the ability to make a tremendous impact on the assimilation of light by plants. Less than 10 per cent of a vine's annual growth above ground results from the action of its roots in the earth. Over 90 per cent comes from the leaves via photosynthesis and the atmosphere. Photosynthesis involves the transformation of invisible, intangible, 'cosmic' matter into matter. Stimulating photosynthesis in the leaf by spraying horn silica 501 helps chlorophyll formation. This in turn stimulates enhanced fruiting (fertilization, seed formation) both during the year the vines are sprayed, and in the following year too because the reserve buds for the following year form on the current year's fruiting shoots. Vines sprayed with horn silica 501 produce grapes with increased disease resistance, enhanced aroma, colour and flavour, with higher nutritional quality (lower nitrates, increased dry matter) and which produce wines with enhanced ageing potential.

Thus horn silica 501 complements horn manure 500 by putting the silica-rich flesh on the horn manure 500's calcium/lime-filled bones. 'Plant life as we know it today can thrive only when these two forces – the forces of substances like lime and like silica – are in equilibrium and are working together properly,'[43] said Steiner, as if the horn manure 500 were 'pushing from below' via the roots in the soil as the horn silica 501 'pulls from above' via the leaves in the atmosphere.[44] The horn preparations thus allow two parallel streams of etheric formative forces to flow. One stream flows downward into the earth and is associated with moisture, soil fertility and digestive processes (horn manure 500). The other stream, flowing upward into the atmosphere, is associated with dryness, fruiting and ripening processes (horn silica 501).

However, it is tempting for biodynamic farmers in general and winegrowers in particular to focus overly on horn manure 500 at the expense of horn silica 501. There may be three reasons for this. The first is that because the organic movement has hammered into farmers the importance of soil health, horn manure 500 is seen as the all-in-one restorative for soils whose mineral balance and microflora have been adversely affected by conventional modern farming techniques. Second, spraying soil with a brownish-coloured 'liquid manure' which horn manure 500 can appear to resemble seems to make more sense to farmers than spraying what resembles glorified water over crops; after all, the contents of a bucket of stirred water and one containing horn silica 501 look essentially the same. Finally, winegrowers – nervy folks who generally get to make wine at most fifty occasions a lifetime – can't always see why attracting warmth forces into their vineyards, which are invariably located in hot and especially sunny places anyway, is really necessary.

The bottom line, says Hugh Lovel, is that if you are bothering to spray horn manure 500 and as a result are activating microbes, nutrients and growth forces in the soil whilst stabilizing its soil nitrogen you're missing a trick by not giving equal importance to horn silica 501 – for it is the horn silica which is going to help drive all that goodness into the crops right as far as their growing tips, preventing plants from getting so lush, weak and watery their crops are poor in terms of yield, quality, disease resistance and potential shelf life. Horn silica helps sap move especially forcefully in spring.[45] For winegrowers, horn silica 501 produces vines with more upright and less floppy shoot growth, shoots and leaves which show greater photosynthetic activity, healthier vines in general and grapes with riper flavours, balanced (meaning lower) sugar levels and enhanced nutritional quality.

Modern wines contain significantly more alcohol than their predecessors. Climate change and the mania (which has now peaked, it seems) for planting cool-climate grapes like Chardonnay and Merlot in unsuitably hot areas where they can quickly over-ripen are two causes. Another is that old wineries lacked temperature control so grapes were picked earlier out of necessity, their lower levels of fermentable sugar (potential alcohol) helping prevent tanks overheating and leading to incomplete fermentations and vinegary wines. But another key reason why contemporary grapes – even from cool-climate varieties grown in cool climates – contain higher sugars is the advent of soluble fertilizers. These feed vines direct rather than via the soil as compost does. Fertilizers promote sugar rather than flavour accumulation, meaning grapes which used to taste ripe at 13.5% ABV now need to be 15% for comparable flavour intensity.[46] In conventional vineyards it is usual to see grapes being picked before a single leaf has fallen, but in biodynamic vineyards grapes should attain flavour ripeness only as the leaves are already starting to fall as nature intended. Horn silica 501 contributes to an alignment of flavour and sugar accumulation, leading to clearer expression of *terroir*.[47]

Hugh Courtney, who ran America's biggest supplier of biodynamic preparations, the Josephine Porter Institute for Applied Bio-Dynamics in Virginia, from 1985 to 2009, having taken over from its eponymous founder, maintains that biodynamic agriculture has been historically weak in the United States due to underuse of both the silica-based preparations, horn silica 501 and common horsetail 508. Courtney says these two 'cosmic force carriers' are especially vital to the American continent, an area of the western hemisphere viewed by Steiner as being dominated by earthly forces. The eastern hemisphere, meaning Asia and Australasia, was more influenced by cosmic forces, thus a greater (if not exclusive) focus there on horn manure 500 might be appropriate, and is the approach taken in Australia by Alex Podolinsky (see Chapter 5, Alex Podolinsky's prepared horn manure 500 + 502–507 spray). 'In my view we have neglected using precisely those preparations that we should most be using in the Western hemisphere,' Courtney concludes.[48]

Making horn silica 501

Having described in the fourth lecture of his *Agriculture* course how horn manure 500 should be made, Rudolf Steiner described how to make horn silica 501. 'Once again, take cow horns, but instead of stuffing them with manure, fill them with quartz that has been ground to a powder and mixed with water to the consistency of a very thin dough.' Quartz, the most common mineral found on earth, is mainly composed of the compound silica. The most familiar silica-rich substances we come across in our everyday lives include sand and glass. In biodynamics, quartz in crystalline form, usually gathered either as rocks from mountain slopes or river beds, is commonly used.

However, Steiner also directed that quartz could be substituted by orthoclase (potassium aluminium silicate), or potash feldspar.[49] Hugh Courtney postulates that while horn silica 501 made from quartz rock crystal may be most effective for those on mainly heavy clay soils, those

*The first stage in making horn silica 501 is pouring the thick paste made
from mixing ground silica with water into the horns*

on sandier soils might benefit from using a form of horn silica 501
with a stronger relationship to the lime end of the lime/silica polarity.[50]
In this case orthoclase or feldspar with their constituents of calcium
or related elements – feldspar being principally aluminium silicates of
potassium, sodium and calcium – would be used.

The rock crystals or pieces of quartz must be broken up into small
pieces and then crushed by being finely ground to the consistency of
flour.[51] The cow horns are filled once the ground silica is mixed with
water to produce a paste which should be just on the point of being
runny.[52] Standing the horns upright or in a sand box allows the horn
openings to be filled level and to the brim, leaving them to settle
overnight. The small quantity of water which will rise to the top can
be poured off so that more of the paste can be added. When the horn's
contents have dried and turned solid, close the horn opening with damp
soil. The filled horns are buried in a pit of similar depth to that used
for horn manure 500. The horn silica pit should not be near any horn
manure pits and should be in the open, so that the sun can reach the site
all day.[53] The same pit for horn silica may be reused again and again, as
is the case with all biodynamic preparation pits.

Freshly filled horn silica 501 horns rest in a sand box to dry off before burial. Guard dog optional.

Horn silica 501 should spend six months in the ground during its preparation, either between the spring and autumn equinoxes – when the earth's forces are outwardly directed, or centrifugal – or between summer and winter solstice. Bouchet recommends burying the horns under an ascending moon and for those also following the sidereal moon when this stands in the flower/light constellations: Waterman in the northern hemisphere, Scales in the southern hemisphere and Twins in either hemisphere.[54] Burying the silica-filled horns over summer allows them to be 'exposed to the summer life of the Earth', said Steiner, so that the silica inside will then form complementary forces to those carried by the 'spiritual manure', his name for horn manure 500.[55] Just as cow horns concentrate the earthly forces in the cow manure for the horn manure 500, so the horns concentrate the solar forces in the silica.

Christian von Wistinghausen says pink spots which are often formed by micro-organisms on the outside of the silica horns are of no concern. Horns for horn silica 501 are only used once (see the note to horn manure above on the disposal of used horns, p.20). Some practitioners feel that the silica carries even stronger forces if the filled silica horns are reburied for several years, instead of the usual single spring/summer burial and autumn/winter lifting.[56]

Storing horn silica 501

Horn silica 501 is stored in a clean, clear glass jar which is covered but not airtight and kept in a position which is sunny, in that it catches the morning sun but is out of direct sunlight.[57] This preparation should never be stored in the dark. It is said to keep indefinitely if dry.[58] Shake the contents of the jar from time to time.[59] The contents of one horn should provide around 125 grams of material, enough for 25 hectares of cropland if sprayed at 5 grams per hectare.[60]

Using horn silica 501

A horn silica spray application for a hectare of land is prepared by adding 5 grams or one level teaspoon to 30 to 70 litres of water and then dynamizing this for one hour in water. The water can be cold, lukewarm or quite warm, in contrast to the water for horn manure 500. No pre-filtration is usually necessary as the silica should be a fine enough powder which will not block spray apparatus. The liquid should be sprayed within three hours of being dynamized. Pierre Masson suggests stirring for horn silica 501 is best initiated at the first signs of dawn, rather than in the pre-dawn dark.[61]

Once dynamized, the horn silica 501 spray is directed skywards at the sun as a fine mist over the tops of the vines. Horn silica 501 is the quickest biodynamic field spray preparation to apply, for even with the lightest of breezes several rows can be sprayed with a single pass. The key practical consideration to be aware of when spraying horn silica 501 is to avoid spraying it when the wind is strong. Pierre Masson recommends a pressure of at least 2 to 3 bars and up to a maximum of 12 for this spray.[62]

Spreading vegetative growth is encouraged if horn silica 501 is sprayed in the early morning, the time when the earth begins to exhale forces, and as the sun is rising, to establish a connection between the crops and the light forces.[63] Ideally morning spray applications should be undertaken on days which are likely to be warm and at least partly sunny.[64] Most winegrowers start spraying horn silica 501 as the morning sun rises over the horizon, early enough for morning dew to be still present on foliage,[65] whose presence is said to be especially beneficial.[66]

Hugh Courtney and freshly excavated horn silica 501 made at the Josephine Porter Institute for Applied Bio-Dynamics

They stop at least an hour before the sun reaches its highest point to prevent any risk of leaf scald. Pierre Masson says a sudden shower soon after spraying will not lessen this preparation's effect.[67] Spraying just before heavy rain is forecast is inadvisable.

Horn silica 501 is generally used on its own rather than being mixed with other sprays. Some growers who add a dose of the common horsetail 508 spray preparation at the end of a horn silica 501 dynamization argue that by combining two sprays with one they are saving time. In fact they may just as likely be wasting it, and for two reasons. On a practical level adding one undynamized liquid, the common horsetail 508 in this case, to another partly or fully dynamized one, the horn silica 501, will nullify the effect the stirring has had on the latter. Those seeking to combine the two preparations should do so before the one-hour dynamizing period is initiated. Even then, the combination of these two preparations in a single spray may create an opposition on a 'forces' level between the pulling-upwards force that horn silica 501 brings to plants, and the downwards-pushing force common horsetail 508 engenders.

Spring applications of horn silica 501 are generally seen as ideal as this is when the seasonal movement of both the sun's arc in the sky and the foliage upon which horn silica 501 works is ever upward. For this reason, spraying when the moon is in its ascending spring–summer phase makes added sense. Winegrowers following the sidereal moon cycle as well spray when the moon stands in the fruit–seed/warmth constellations: Ram in the northern hemisphere, Lion in the southern hemisphere or Archer in either hemisphere. However, few winegrowers lose sleep if forced to spray horn silica 501 under either the root/earth, flower/light constellations or even leaf/water constellations.

Biodynamic winegrowers generally spray horn silica 501 two or three times annually. Four or more sprays per season is considered unusual. Key stages in the vine's vegetative growth cycle at which horn silica 501 might be sprayed include:

- when the first five leaves have appeared in spring;
- just before flowering commences;
- after the flowers have been fertilized, when the embryonic bunches become visible;
- between fruit set and *véraison* (see below), if wet summer weather has augmented the risk of fungal disease;[68]
- when the grapes begin changing colour in August (*véraison*) to speed this process up;
- just before harvest;
- just before leaf fall.

Although assimilation in plants starts as soon as the first green leaves appear, spraying horn silica 501 at this very early stage will inhibit shoot and stem growth (cell division) due to a lack of sufficient foliage.[69] Winegrowers usually make their first application of horn silica 501 from the five-leaf stage onwards, when that part of the plant from which the harvest will be gleaned is beginning to develop.[70] Spraying horn silica 501 when crop flowers are actually open may result in poor seed setting.[71]

Spraying horn silica 501 at Millton Vineyard, Gisborne, New Zealand

Andrew Lorand says that what matters for both biodynamic horn sprays is regularity of use.[72] He advises spraying horn silica 501 during the last three or four months of the growing season and in the fourteen-day periods before new moon, because these seasonal and lunar periods are when 'tightening' siliceous forces are strongest (to put this in context, see p.23 for his view on when horn manure 500 should be sprayed).

In wet, cloudy or cool years, when grape ripening is slow or looks like slowing down, horn silica 501 can be sprayed from between when the berries change colour – *véraison* – until harvest. In this case, rather than being sprayed in the morning as it is in spring, horn silica 501 is sprayed in the afternoon. This induces an 'autumn mode' to the vineyard during the period when afternoons are getting shorter, telling the vines to hurry up and ripen their grape skins, sugars and pips or seeds. Spraying horn silica 501 in the afternoon is said to aid senescence by encouraging sap to flow down into the roots. This makes the top half of the plant less 'watery' which aids seed and grape ripening, and flavour development and retention.

Hugh Lovel says that when an afternoon application of horn manure 500 has been made, horn silica 501 may be sprayed the very next morning, a form of sequential spraying (see Chapter 5, Hugh Courtney's sequential spray technique) which has 'a wonderful balancing effect on the land and atmosphere'.[73] Like horn manure 500, horn silica 501 is not generally applied during the height of summer.

Biodynamic growers see the time between the end of grape picking and the first winter frosts which bring leaf fall as another opportunity to spray horn silica 501. During this period the leaves are still photosynthetically active and food in the form of carbohydrates can be sent to the vine's trunk and roots. Rudolf Steiner said that autumn rather than spring was the most truly fertile moment of the year because this is when the sun sends its forces in the form of nutritious sap to feed the earth via plant roots and the micro-organisms living upon them.[74] Spraying horn silica 501 over vines now allows them a final burst of photosynthetic activity, to make optimum use of diminishing levels of sunlight because between autumn equinox and winter solstice the earth spends more time in darkness than light each day. By making the top half of the plant less 'watery' an added benefit of afternoon sprayings of horn silica 501 now is that openings left on vine shoots by falling leaf petioles can close or tighten more quickly. This allows vines to enter winter dormancy fully sealed and protected. Autumn applications of horn silica 501 thus allow weaker vines to start the following season in better shape, and to produce riper canes the following year.

Spraying horn silica 501 on crops whose plant roots are not well established – young vines or vines which have yet to reach bearing age – may produce unwanted side effects like inhibited shoot and stem growth.[75] The risk can be nullified or at least reduced if the young vines have been planted in properly rested, biodynamically composted soil which has also been sprayed with horn manure 500. It may also pay to follow Proctor's recommendation of spraying horn silica 501 during an ascending moon phase and when the moon is also in opposition to Saturn (see Chapter 7). This lunar window occurs for only one day a month, but not every month – so while it may be unsuitable for the entirety of the biggest vineyard holdings it can certainly be used for the youngest vines which typically make up less than 5 per cent of the total of any wine operation at any given time.

Common horsetail 508

This spray preparation lacks the cornerstone status of the two horn (manure and silica) preparations 500 and 501, and shares none of the near-mystically weird methodology of the solid compost preparations 502–506. It doesn't get sheathed in an animal organ, buried in the ground or hung in the sun. There's not even any explicit need to wait for the right type of moon to prepare it or even to stir it before using it, although dynamizing it for twenty minutes to an hour will do no harm (see Chapter 4 for why). Perhaps because it doesn't *feel* like it's a real biodynamic preparation, plenty of biodynamic winegrowers, especially those in the Antipodes and north America ignore it (see p.27). Maybe they feel that because common horsetail must be boiled or macerated in water to liberate its silicates this means that its action on farms and crops is more physical (substances) than (bio)dynamic (forces). Those growers who ignore common horsetail 508 risk wondering why the 'eight rather than nine' biodynamic preparation road hasn't taken them as far as they'd hoped in terms of controlling weeds, pests and especially fungal diseases. Yet this is precisely what common horsetail 508 helps to do, not by acting directly like weedkillers, pesticides or fungicides do but indirectly – by reining in the excessively strong moon forces which encouraged the weeds, pests and diseases to establish and reproduce in the first place.[76] Common horsetail 508 is not, however, some miracle fungal-disease cure but a prophylactic spray with a mild fungal-disease-suppressing effect. It is most effective when used in conjunction with the other eight biodynamic preparations.[77]

The common horsetail (*Equisetum arvense*) plant is a very light, fern-like perennial which grows along shady riverbanks and lakesides, or anywhere where drainage is poor.[78] The horsetail family is large but – unlike marsh, field, wood, tall and other horsetails – common horsetail grows from fungal spores, rather than sexually produced seeds. That's what makes it so special: it's a green plant that has overcome its fungal origins and stays free of disease in fungus-dominated places by existing in a purely vegetative state.[79] The spores produce rhizomes in the soil which in spring produce the bare stalks that grow into what look like sparsely branched mini Christmas trees. To make the preparation, the plant's green fronds (not its rhizomes) are harvested before midsummer or at summer solstice at the latest.[80] This is when the fronds' silica content is highest, rendering them especially hard and firm.[81] The fronds resemble elongated but less prickly pine needles; more like a bottle brush than a horse's tail in fact.[82] They contain the highest concentration of silica (over 75 per cent) in the plant kingdom and appear so luminous they could almost be nature's equivalent of a light sabre. It is the silica which gives common horsetail 508 its drying or 'tightening' effect, like the silica sachets at the bottom of the box your new camera or pair of binoculars came in.

Despite already having one silica-based preparation in horn silica 501, biodynamicists do need the second one in the form of horsetail. Whereas horn silica 501 encourages upward movement, to get the crop to express itself vertically by connecting with the light and heat forces of the sun, common horsetail 508's effect is downwards, as its silica harnesses sun forces in such a way as to push fungal diseases down off crops and back into the soil where they belong. Steiner said this preparation should be used 'when the moon's influence [which affects the fluid element] is too strong [and] the soil is overly enlivened [and] vitality works up too strongly from below ... because of the effect of the moon, there is insufficient force for seed formation [needed for healthy reproduction and ripening] ... as a result, the seed, or the upper part of the plant becomes a kind of soil for other organisms. Parasites and all kind of fungi [which really belong lower down in the soil] appear [instead, above the soil on crop plants].'[83] Hence biodynamicists see that fungal disease parasites are more likely to jump from the soil onto vines if water-soluble mineral fertilizers designed to act only on the soil's watery element rather than on the soil itself are used instead of solid compost. Andrew Lorand finds common horsetail especially effective at driving fungal disease organisms back to their as-nature-intended soil realm when sprayed in the week before new moon.[84] This lunar phase itself brings a tightening, cleansing effect whilst also aiding photosynthesis and thus fruiting, ripening and the keeping qualities of grapes (wine).

Note, however, that as common horsetail (*Equisetum arvense*) is considered an invasive species in parts of Australia and throughout New Zealand, alternatives such as she-oak (*Casuarina stricta*) or the Australian pine (*Casuarina equisetifolia*) may be used instead, in the latter's case from a male tree only.[85] In Chile and Argentina *E. gigantum* may be used.

The common horsetail 508 spray preparation can be made either as a fresh tea or as a fermented liquid manure.[86] Pierre Masson suggests harvesting fresh plant material each year to maintain this preparation's effectiveness.[87]

Making common horsetail 508 as a fresh tea

The plant fronds are picked and dried in airy shade until needed, but should retain their green colour.[88] For one hectare of vines, add roughly 100 to 300 grams to 2 to 5 litres of water. This is then brought to a rolling boil on a low flame to 'loosen up'[89] the fronds, then covered and left to simmer for thirty minutes to an hour. The resultant tea concentrate will have a pale yellow-green or brown colour.[90] From the concentrate, make a 10 to 20 per cent dilution either by adding water or by adding the concentrate directly to vineyard sprays like those used to combat downy and powdery mildew. The common horsetail tea can also, of course, be sprayed on its own directly on the vine leaves and even on the flowers during bloom, as an extra fine mist.[91] It can be sprayed at two-week intervals, around full moon or lunar perigee to rein in potentially overmighty lunar growth forces.

Making common horsetail 508 as a fermented liquid manure

Common horsetail tea left to macerate in a glazed crock or other storage container with a loose fitting lid for ten to fourteen days at ambient temperature or in the sun will ferment,[92] with no negative effect on its usefulness.[93] A distinctive-smelling gas (hydrogen sulphide) is given off, sulphur being one important constituent of common horsetail, and silica being another. This gives fermented horsetail a reinforcing effect on vineyard sulphur sprays when combined with them. Once fermented, the liquid can be strained and stored ready for later dilution (to 20 per cent), or topped up like a 'mother' to which fresh tea is added whenever fermented concentrate is drawn off. The fermented liquid-manure version of common horsetail is usually employed as a soil rather than crop spray, to carry solar and other warmth/light forces into the soil. It is said to be most effective if pre-stirred for between twenty minutes and an hour,[94] and then sprayed on soil which is moist – and sprayed on its own rather than being mixed with another.[95] Thus it can be sprayed in less-pressured moments, such as late autumn and early spring on areas of the vineyard where experience suggests potentially harmful fungi are likely to appear.[96]

SIX COMPOST PREPARATIONS 502–507

Rudolf Steiner created the six biodynamic preparations for compost piles – yarrow 502, chamomile 503, stinging nettle 504, oak bark 505, dandelion 506 and valerian 507 – as a response to modern methods of fertilizing which 'sometimes give astonishing looking results' but risk ultimately turning potentially top-quality crops 'into mere stomach-fillers. They will no longer have real nutritive power for human beings. It is important not to be deceived by things that look big and swollen [a hydroponically grown tomato, perhaps?]; what is important is that their appearance be consistent with real nutritive power', meaning food which could nourish both body and spirit. [97]

Steiner said that growing sound and substantial plants would not be achieved by enlivening the watery part of the soil via mineral (water-soluble) fertilizers, 'because no further vitalization proceeds from the water that seeps through the soil. We have to enliven the soil directly, and this cannot be done with mineral fertilizers, but only by means of organic material that has been conditioned to [physically] organize and [spiritually] enliven the solid earth itself. To indicate how this stimulus can be imparted to manure or liquid manure, or any other sort of organic matter, is the task of spiritual science [anthroposophy] with respect to agriculture … of infusing the manure with living forces, which are much more important to the [crop] plants than the material forces, the mere substance.' [98]

Therefore, as regulators rather than inoculants or activators, the six biodynamic compost preparations' main role is to radiate intangible forces in the compost rather than to supply physical material, like minerals. The compost preparations act as foci through which various forces and influences can work into the different types of organic matter being composted, guiding and supporting their breakdown to create a harmonious whole which, once spread on the land, mobilizes substances and makes crop plants more receptive to both these physical substances and to intangible forces streaming in from the cosmic environment. [99]

Steiner said making the biodynamic compost preparations 502–507 does take 'a certain amount of work … but *if you stop and think* [my italics] … it actually takes less work than all the fooling around in chemical laboratories that goes on in the name of agriculture, and which also has to be paid for somehow. You will find that what we have discussed is much more economical.'[100]

In other words, either stimulate long-term fertility and life force in the soil by learning to make compost and the fiddly compost preparations that go with it, or choose off the shelf, soluble fertilizers whose value for money is based in part on the fact that those aquatic organisms affected by the dead zone in the Gulf of Mexico, for example (see p.8), which these fertilizers have caused, are unlikely ever to file a lawsuit against the manufacturers.

Storing compost preparations – dry or moist?

The biodynamic compost preparations require careful storage. Each one is given its own separate storage container (see photo). Proctor advises keeping the preparations 'cool and moist (not wet) in glazed [or glass] jars with loose-fitting lids in a peat-lined box or just surrounded by peat. *Never* put them in a refrigerator. They are best in a cool shady place, for example

Storing biodynamic preparations in glass jars in individual copper containers in peat-lined wooden boxes keeps them moist and humic

under a house or in a cellar.'[101] Peat is said to block or at least minimize any antagonistic forces (radiation, electrical currents) that may harm the etheric formative forces which the preparations carry. The preparations tend to dry out in unglazed clay vessels, such as flower pots.[102]

Whether the five solid biodynamic compost preparations 502–506 should be stored moist or dried is a matter for debate. Drier preparations were favoured by the main producer of biodynamic preparations in Germany, Christian von Wistinghausen, who died in his seventies in 2008. He told me he preferred 'dry preparations and since pharmacies have been drying their herbs for millennia in order to store them, it cannot do much harm. By keeping them in a moist state they are already starting a breakdown and composting process of their own. I want them to do this in my pile of compost, not before. So I keep my preparations dry, and thus alive, like dry seeds, drying them using only air and shade. When it is time for the preparations to begin their work in the newly made compost pile this will happen naturally because as soon as the preparations are inserted into the piles they will gain the moisture they need.' This approach means that as the preparations are then no longer in their living state but have become dormant, they do not turn to earth after a few months like those held in a moist state can do.[103]

However, Alex Podolinsky (see Chapter 5, Alex Podolinsky's prepared horn manure 500 + 502–507 Spray) is critical of dry biodynamic preparations, calling them lifeless and 'untransformed', meaning they had neither been prepared nor stored properly and had thus not undergone the 'transubstantiation' process necessary to produce the high-quality crops biodynamics should be renowned for.[104] Podolinsky favours preparations which are moist and humic and which remain so when stored in jars in individual copper containers in peat-lined wooden boxes.

How the compost preparations are inserted into compost piles and the amounts required is covered below.

Yarrow 502

Rudolf Steiner said yarrow has a beneficial effect simply by its mere presence.[105] Although in the words of one Antipodean biodynamic vineyard consultant, yarrow can be 'a tough little blighter to get rid of when sown as part of a vineyard cover crop mix', Steiner said that because yarrow is not actually harmful you should never try to get rid of it.

Yarrow (*Achillea millefolium*) is a member of the daisy (*Compositae*) family, a hardy, persistent and perennial herb often considered a weed but sometimes sown as a vineyard cover crop.[106] It grows wild and up to waist height in Europe and North America in patches along roadsides and in pastures and meadows. The flowers, small and finely divided, are white, creamy yellow or pink. They form in flat-topped clusters 'like a differentiated white or pink mirror that faces upwards. The leaves are most beautifully and finely divided, which is why the ancients called yarrow *supercilium veneris*, the eyebrow of Venus.'[107]

Despite being a 'tough little blighter' yarrow's roots are shallow, as Proctor points out[108] and 'mainly at the surface: they are not deeply attached to the earth … Each flower is like a little chalice – a receptacle for receiving the beneficence of the cosmos. This connection with the cosmic forces, one imagines, enables the yarrow to concentrate many trace elements. It is a wonderful example of … plant dynamics. Yarrow has been found to contain a measurable amount of potassium and selenium even when the soil in which it grows lacks these minerals. In 1985 in Reporoa, New Zealand, liquid manure was made from yarrow plants growing on land where soil tests had shown deficiency of potassium and total absence of selenium. Analysis showed the liquid manure to contain measurable amounts of these minerals.' Hugh Lovel says yarrow 502 is one of the best biodynamic preparations for detoxifying the soil.[109]

Steiner called yarrow 'a miracle of creation … as if some plant-designer had used an ideal model in bringing sulfur into relationship with the other plant substances',[110] because he saw yarrow's highly dilute sulphur content combined with potassium in such an ideal way. Yarrow's strong relationship with potassium is shown by its firm stalks. As a regulator, yarrow draws in fine dilutions of substances from the celestial sphere, using potassium and sulphur to bring cleansing (light) forces to the soil[111] and thus into the organic processes of the farm.[112] While each plant used to make the six biodynamic compost preparations 502–507 has a good connection with sulphur, as seen by their well-cut and deeply lobed leaves,[113] Steiner said yarrow's relationship with sulphur is 'truly exemplary'. Sulphur is what spiritual forces need to be able to carry organic substances like carbon, nitrogen and so on around the farm organism, into the soil, the crops and thus into our food. Steiner said that by using the yarrow preparation in our compost, animals and

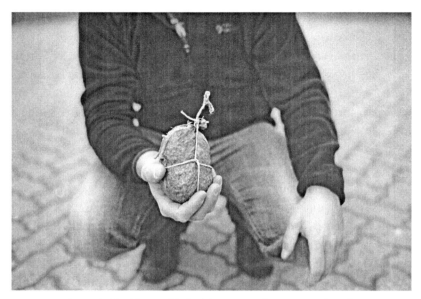

*A stag's bladder stuffed with yarrow flowers ready to be
hung in the sun over summer*

humans will become more sensitive to their surroundings and 'any
weakness of the astral body', meaning the spirit's ability to sense and
carry forces, will be rectified.

Making the yarrow 502 compost preparation

The yarrow flowers are collected in sunny weather but only when all the
florets in the cluster have opened.[114] The easiest way is to cut individual
florets, leaving some stem attached, as it makes handling them a little
easier. The hard stems can be clipped off later to prevent them piercing the
stag's bladder they will eventually be placed in. Ideally the flowers are used
fresh, but in Europe and most other parts of the wine world – Argentina
being an exception – where yarrow flowers between late spring and early
summer, the period during which this preparation should hang in its
animal sheath in the summer sun will then be too short. Thus the florets
will need to be picked in summer or autumn and dried ready for use the
following spring.

The florets can be dried on hessian sacks or on slatted trays, or
can be hung in bunches in an airy room as for herb teas. Bouchet
recommends harvesting the florets when the sidereal moon stands in

the fruit–seed/warmth constellations Ram, Lion or Archer and ideally when the summer sun also stands in Lion.[115] From Bouchet's northern-hemisphere perspective this currently means between 11 August and 15 September. In the southern hemisphere you would, by implication, wait for the sun to stand in either Ram, from 19 April to 12 May, or Archer, from 21 December to 16 January.

When Steiner talked of using yarrow 'in the right way in our compost' he meant the florets had to be prepared by being stuffed into the bladder of a red deer stag (*Cervus elaphus*) and remain there for six months.[116] Steiner noted how the stag's antlers share a similar form to that of the yarrow flower stems, and how, like yarrow, the stag forms a particularly strong connection to the celestial sphere – but through its antlers.

The stag's antlers are made of bone excreted from its interior, sense antennae extending towards the celestial pole through which the stag releases excess forces.[117] The antlers also sense forces which radiate into the stag's metabolic processes and into the bladder.[118] This organ is often emptied when the senses heighten in moments of fear, surprise or

The filled stags' bladders for the yarrow 502 preparation are exposed to the sun over summer. Leaving plenty of space around them prevents damage when blown by the wind

ecstasy. The bladder is the focal point for everything the animal senses. Yarrow is used medicinally for bladder and kidney ailments, cleansing and purifying, eliminating salts dissolved in the urine, and above all for nitrogen and potassium.

The qualities Steiner said he identified in yarrow are 'especially strongly preserved in the bodies of humans and animals by means of the process that takes place between the kidneys and the bladder. As thin as it may be in terms of substance, in terms of its forces a deer bladder is almost a replica of the cosmos.'[119]

Steiner said that putting yarrow into the deer bladder 'significantly enhance[s] its inherent ability to combine sulfur with other substances'.[120] In Steiner-speak sulphur is the element which – by carrying the spiritual force or will – organizes how matter (e.g. the carbon plants absorb from the atmosphere which becomes carbon-rich compost, and the key macro-nutrient potassium) is formed, shaped, sculpted. If any preparation could be considered a microcosm of biodynamics, yarrow is it.

Stag bladders can be sourced either via local game wardens, hunters or abbatoirs, or even from fresh road kill. Ideally the bladder will come with the male reproductive organs intact, so there is no chance it could have come from a female, and unrinsed. The urine still inside can be kept in a glass container and used later to rinse the bladder before it is stuffed with the yarrow florets.[121] Unless it is to be used fresh, the bladder can be inflated by removing part of the male organ and using the still-attached part of the urinary tract as an entry hole through which air can be blown or pumped. Once inflated the bladder can be tied off with string and hung for drying, and stored until needed for stuffing.

Before being stuffed, any bladders which had been stored dried or frozen will need softening with a dilute and warm rather than freshly boiled tea made from yarrow (the stems and leaves rather than flowers will do for this). The same tea can also be used to moisten the florets to make the stuffing operation less likely to result in a split bladder. Once the bladder is softened it can be rinsed with any saved animal urine. Making a small cut about two fingers wide in the side of the bladder near the urinary tract allows the bladder to be stuffed with the moistened flowers, using a wooden dowel if necessary. Once filled the sheath should have the firm consistency of a softball[122] and should have retained its natural spherical form.[123] After filling the sheath the opening is then sewn up and the bladder is ready to be hung outside. It should be waterproof so the flowers inside will not get mouldy in wet weather.[124] Bladders which are too tightly filled and split when they dry out can be patched up with a new piece of bladder skin.

In spring the filled bladders are hung in a dry place exposed to the sun (facing the equator) and remain there until autumn when they are buried. The aim is to expose the filled bladders to the same weather conditions as will affect the farm during spring and summer.[125] Reflecting its surroundings during its making allows the yarrow compost preparation to regulate those same surroundings via the compost when this is spread on the soil. Typically bladders are hung in trees, under the eaves of buildings or on purpose-built stands. The bladders must remain intact and should be hung far enough apart so as not to bang against one another (or anything else) in the wind. They must be out of the reach of dogs and other animals, and any bits of fat must be scraped or picked off to discourage pecking from birds. Netting can be used, leaving plenty of space.[126]

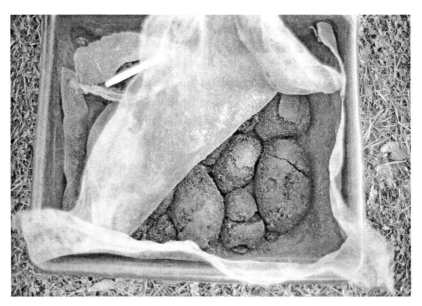

Freshly excavated yarrow 502 bladders

After spending a full summer outside the filled bladders should be taken down in autumn from their hanging place and buried for the duration of winter. This is the period when the earth opens up to celestial forces as it inhales. The bladders should be buried at the same time as the horn manure 500 preparation and in a pit of similar depth, i.e. covered by around 30 to 45 centimetres of soil. The bladders should be a slight distance apart so that each one is both surrounded and covered by soil. Before backfilling the pit, placing a layer of sacks made from natural (residue-free) fibres over the top of it helps mark the boundary between the covering soil and the bladders when the pit is opened again.[127] Burying each bladder in a small bottomless wooden box surrounded with earth also helps act as a marker. In tropical countries termites will destroy the wood so an unglazed earthenware pot may be used instead. This is buried in a brick-lined hole, again making sure the preparation is surrounded by soil.[128] Protect the yarrow bladders from dogs, badgers, foxes and other large burrowing animals by laying roof slats or metal screening (gauze) over burial pits. The metal should be copper, which enhances yarrow's relationship to Venus, rather than iron which is related to Mars and which, by being opposite in polarity, will inhibit it. Lining the bottom, sides and top of the pit with freshly cut elder branches deters mice who find the odour repulsive.[129]

The filled bladders should be dug up in late spring, or just after Midsummer Day at the latest.[130] Bouchet suggests a day when the planet Mercury stands in the fruit–seed/warmth constellation Ram.[131] The bladder will have almost or completely rotted away but the yarrow flowers should have held the bladder's shape. Care must therefore be taken to keep the preparation free of excess soil by using a blunt-edged implement like a wooden spatula when scraping away any remaining bladder membrane.

Storing the yarrow 502 compost preparation

The excavated yarrow preparation should be light brown in colour and feathery light in texture.[132] Proctor notes that as the yarrow preparation never actually forms into humus – one can still see the structure of the flowers in it (see photo, p.45) – and its nature is quite loose, it can dry out rather quickly.[133] 'Dry preparations will lose their strength', Proctor says, and recommends placing the preparation in a glass jar with a loose-fitting lid and then placing this is in the designated (peat-lined) storage box with the other compost preparations. To help maintain its moisture the preparation can be pressed down a little to firm it in the jar. A teaspoon or two of water can be added if it looks too dry.

Using the yarrow 502 compost preparation

Steiner stressed just how strong the 'radiant energy' the yarrow preparation gives off will be, even in proportions which are tiny compared to the volumes of compost to be treated.[134] Yarrow, he said, 'is able to radiate its effects through large masses of manure because its own highly dilute sulfur content is combined with potassium in such an ideal way'.

The contents of a single stag's bladder should produce enough yarrow preparation to treat 250 hectares of cropland.[135]

Chamomile 503

Both chamomile (tea bags) and sausages (pork, beef, tofu) feature in almost every contemporary kitchen, yet the notion of combining both in a chamomile sausage, which is what this preparation is, seems anathema. But having made chamomile sausages for the first time, you might conclude that although biodynamics is bizarre, modern life is even more so because so few of us in our blithely 'disconnected from the food we eat' world have ever had to make any kind of sausage before, meat, tofu, chamomile or otherwise. Jeff Poppen sums up why Rudolf Steiner created these floral sausages: 'Chamomile flowers make a soothing and relaxing tea that is good for stomach aches. They also have sulfur and calcium in just the right proportion and quality to make a preparation which gives manure [being composted] the power to receive enough life into itself so that it can transmit it back to the soil.'[136]

What Steiner said in more detail was, 'If we want to give our manure the possibility of adsorbing vitality that it then can impart to the soil where the plants are growing, we must make the manure especially able to combine the elements that are necessary for plant growth. In addition to potash, these include calcium, various compounds of calcium. In the case of [the] yarrow [502 compost preparation described immediately above], we were dealing primarily with the effects of potash; if we want to draw in the effects of calcium as well, we need another plant [chamomile]. This plant may not arouse our enthusiasm as readily as yarrow, but it too contains homeopathic sulfur, which enables it to attract the other substances it needs and incorporate them into an organic process … We may not simply say that chamomile is characterized by a high content of potash and calcium. Rather, the situation is as follows: Yarrow develops its sulfur forces primarily in the potash-forming process; it therefore contains sulfur in exactly the amount needed for working on potash. Chamomile, on the other hand, works on calcium in addition to potash, and thereby develops what can help to ward off the harmful effects of fructification [fruiting] and keep the plant healthy. So, the marvellous thing is that like yarrow, chamomile contains some sulfur, but in a different quantity because it has to work on calcium too.'[137]

Steiner said that the chamomile 503 preparation helps compost to have a more stable nitrogen content. Its effect is to enliven the soil in such a way that plant growth will be stimulated and more vital. Chamomile's ability to rein in nitrogen, to 'hold in the life forces'[138] seems arbitrary until one understands that before refrigerators were invented chamomile was used to preserve meat, preventing putrefaction and the forces of decay. The cow's intestine performs the same role, holding in the life forces, allowing her digestive system to produce sweet-smelling manure that engenders healthy soil and crop growth. A pile of manure-based compost smells of putrefaction when nitrogen is being lost in the form of ammonia. Chamomile's calming effect, particularly for digestion in humans, is well known and analagous to Steiner's belief that compost treated with the chamomile 503 preparation is capable of producing 'healthier [crop] plants – really much healthier plants'. And healthy plants are the basis for the next cycle of growth to continue.

Making the chamomile 503 compost preparation

Although several varieties of chamomile exist, the flowers needed for this compost preparation should come from true or German chamomile or *Matricaria chamomilla*, also referred to botanically as *Chamomilla recutita* and *Matricaria recutita*. This is the same strain of chamomile found in tea bags, and is quite different from the more bitter-tasting Roman or Italian chamomile or *Chamaemelum nobile*, which is also known as stinking mayweed.[139] The desired (German) chamomile is identified by cutting open the cone on which the little yellow flowers are found. Its inside should be

Trim fat from fresh cow intestines before filling them to make the
chamomile 503 preparation

hollow, whereas that of the Roman chamomile is solid and flatter in profile.[140]

Chamomile is a member of the daisy family. It is an annual plant which dies off after midsummer, soon after it has set and dispersed its seeds. Proctor says 'it has a mercurial quality and tends to pop up in different places each year'.[141] Proctor's use of the term mercurial in relation to chamomile is quite deliberate, because both chamomile and the planet Mercury are related to the intestines and their functions of digestion and assimilation;[142] and a cow intestine – sourced from BSE-free countries[143] – is the designated animal sheath for this preparation.

If relatively large quantities of chamomile are needed it is best to start picking the flowers as soon as flowering begins, which is towards the end of May/early June in the northern hemisphere. Choose a sunny morning before 10 a.m., when the flower petals are horizontal and are in the process of opening. Early morning dew will make the petals hang downwards. Only young flowers should be picked: their centres are slightly pyramidal and greenish and they have a ring of pollen-bearing stamens at the base of the cone. When the cone is spherical and yellow, the flowers are too old to be of much value.[144] Bouchet recommends

Chamomile 503 sausages (intestines) ready for hanging, then burial

harvesting the florets when the sidereal moon stands in any of the flower/light constellations (Twins, Scales or Waterman).[145]

Chamomile flowers which are to be stuffed immediately into fresh cow intestines for hanging over summer before burial should be allowed to wilt slightly first.[146] If the flowers are to be used in intestines which will be both filled and buried in autumn then it is recommended to dry them, as moist flowers are likely to turn mouldy when stored.[147] To avoid this spread the freshly picked flowers on a hessian frame in a warm, shady, airy place but not in direct sunlight, and move them daily until they look dry. The dried flowers can be hung in small sacks that will let the air through, or stored in brown paper bags or glass jars in an airy cupboard. 'Well-dried chamomile flowers have a beautiful fragrance ... [but] watch that mice or moths don't spoil them', Proctor says.

The cow intestines should be fresh, ideally from animals reared on your own agricultural holding. The part required is the small intestine (jejunum and ileum) although the first, spiral part of the large intestine, the colon, can also be used.[148] Running the intestines between forefinger and thumb will expel some of the partially digested contents without emptying them completely or needing to turn them inside out. Fresh intestines can be stored for later use by inflating them with air using a pump, tying them off and drying them by hanging under cover or in the sun.[149] Store the dried intestines in an insect-proof container, perhaps adding a small amount of volatile oil into the container to keep away moths.

The intestine is long and slippery to handle, so cutting sections into 25- to 40-centimetre lengths can make stuffing them easier. Excess fat can be trimmed off using a razor. Any inadvertent nicks can be patched up later. Tying off using a loop at one end of the intestine makes subsequent hanging easier. Usually the flowers are pushed into the other, open end using a wide-stemmed funnel with the help of a wooden dowel; lightly moistening the flowers with chamomile tea makes stuffing easier. The intestines should be tightly stuffed rather than filled to bursting. During the early stages of filling, placing a thumb behind the base of the tied end of the intestine supports it and stops it from splitting under the pressure of the flower heads being forced in at the other end. 'When the intestine is full, tie the end with string. You will notice that any unpleasant smell associated with ageing intestines and warm weather quickly disappears when they come in contact with the chamomile flowers,' says Proctor.[150]

The filled cow intestines are usually buried in autumn, but in handwritten notes for his 1924 *Agriculture* course Steiner suggested that both the chamomile and dandelion (see below) compost preparations would benefit from spending six months hanging in a tree before being buried in the ground, in a similar way to the stag's bladder for the yarrow 502 preparation (see photo, p.46).[151] This would enable the preparation to be exposed to the sun for at least part of the summer. On a practical level, hanging the freshly filled sausages to dry for a few days anyway makes sense, as this helps the intestinal wall to close up firmly around the contents. If the intestines are buried when still fresh they may rot prematurely, and the smell of decomposition will attract predators.[152]

Steiner said that in order for the chamomile to be 'worked on by a vitality that is as closely related to the earthy element as possible' the chamomile sausages 'should be placed not too deeply in soil that is as rich as possible in humus' and 'in a spot that will remain covered with snow for a long time, and where this snow will be shone upon by the sun as much as possible, so that the cosmic-astral influences will work down into the soil where the sausages are buried'.[153] Choose an open (i.e. sunny) site in a field where snowdrifts tend to form and where this snow remains for as long as possible even after winter comes to an end, shovelling a pile of snow over the pit if necessary.[154]

As dogs and foxes will smell decomposing matter, avoid placing the freshly stuffed chamomile sausages by the edge of the pit when burying them.[155] Protect burial pits from burrowing animals in the same way as described for the yarrow 502 preparation, above.

Steiner directed that this preparation be dug up around spring equinox. Chamomile sausages buried the previous autumn rather than the previous spring may wait another month or two. Chamomile sausages lifted in early spring are less likely to be incorporated into the surrounding soil by earthworms.[156] The biodynamic consensus is that the chamomile preparation is often in short supply. It is the most difficult to recover as it breaks down completely and is not easy to locate. Proctor suggests the finished preparation is more easily located if the fresh chamomile sausages are buried in an unglazed earthenware flowerpot, 30 centimetres in diameter, having first surrounded the sausages in the pot with soil. Unglazed pots allow all the earth processes to travel through to the sausages. Alternatively Proctor says one can put the sausages in a 'sandwich' of sphagnum moss in the soil and mark its

Freshly excavated chamomile 503 intestines

position with pegs. When digging up the preparation any soil attached to the intestines should be carefully brushed off before they are slit open, lengthways, so their contents can be removed. Crumble the contents up and dry them as quickly as possible in an airy, shaded place. Any intestine residues or bits of chamomile or soil adhering to them can be chopped up and added to manure piles or liquid manures.[157]

Storing the chamomile 503 compost preparation

When made correctly, this will be the most well-composted of the five solid compost preparations. Although the plant material will partly decompose it will not lose its characteristic scent,[158] even if the scent of the flowers will have become 'more animal'.[159] This preparation can be stored in the same way as yarrow.

Using the chamomile 503 compost preparation

Roughly ten units of chamomile flowers will produce just a single unit of the finished preparation, a single unit being enough for 7 to 10 tons of compost. A 30-centimetre long chamomile sausage should produce enough of this compost preparation to treat 100 hectares of cropland.[160]

Stinging nettle 504

Stinging nettle is the easiest of the six compost preparations to make, even if you'll need to collect great armfuls of nettles just to make a handful of preparation, and do so when the nettle's stings are at their most potent, when the midsummer sun is at its most sweat-inducing and the plant's pollen at its most sneeze-inducing. Stinging nettle is the only one of the five solid compost preparations 502–506 to be prepared without being encased in an animal sheath as a sense organ.[161] Steiner said that a plant as special as stinging nettle just needs to spend some time in a simple hole in the ground for one year for it to be able to infuse the soil with the most potent force of all: intelligence.

Joly likens stinging nettle to 'a sensitive and experienced diplomat' because it 'directs towards its centre what many other plants express at their extremities'.[162] The roots remain close to the surface rather than go very deep into the soil and flowers emerge across the central length of the plant rather than just emerging from the top. Both flowers and the green seeds they produce are very discreet. Joly says that this shows how the nettle 'concentrates on its leaves: their transparent shimmer, almost floral in nature, and their burning sting [in the form of formic acid], are qualities which ought not to be there in the normal run of things … the leaf is the plant's mediating organ between its roots at one extremity and its fruits at the other'.

Steiner held that stinging nettle and other plants which sting, barb or exudate already have enough 'astrality' or inherent ability to sense and carry forces of their own for no animal sheath or sense organ to be needed when making the preparation.[163] As a sheath the earth will do; and earth, particularly waste ground firmed and hardened by an underlying iron pan, is what stinging nettle is mightily adept at dealing with, converting it into something that smells healthily rich in humus. Nettles are in fact so fond of iron, said Steiner 'that they draw it out of the soil and into themselves, and although this does not get rid of the iron as such, it at least undermines its effect on the growth of other plants … [thus] the mere presence of stinging nettle can already be of significance for the plant growth in its surroundings'.[164] Stinging nettle in either its wild form or as a compost preparation and even as a soil spray (see Chapter 6) has many uses for soil: it warms (in the sense of softens) its appearance, balances excesses of both iron and nitrogen,[165] gets otherwise-blocked soil minerals moving, activates enzymes beneficial for crop roots, makes heavy soils more porous, builds humus and prepares soil for nitrogen fixation.[166]

Making the stinging nettle 504 compost preparation

While Steiner said substitutes could be found for other preparation plants, specifically yarrow and chamomile,[167] stinging nettle is one plant 'whose beneficial influence on the manure [compost] is such that it would be next to impossible to find a substitute for it. We are often not very fond of this plant, at least not in the sense of wanting to fondle it … But stinging nettle is in fact the greatest benefactor of plant growth, and it can hardly be replaced by any other plant. If it is not available locally, you really must get the dried herb from somewhere else.' When asked if he was talking about the annual or perennial stinging nettle Steiner answered '*Urtica dioica*',[168] meaning the perennial, European stinging nettle rather than an annual with less sting, like *U. urens,* for example.

The *Urtica* part of its Latin name comes from *urere,* to burn or sting, while the term *dioica* is derived from dioecious (from the Latin for 'two houses'), which is a botanical designation for plants having both male and female reproductive organs borne on separate individuals of the same species.[169] When collecting the stinging nettles, Steiner said to collect the whole plant when it is in flower.[170] The flowers, leaves and stems can be used to make the preparation – but not the roots. Those following the sidereal moon cycle harvest nettles when the moon is in the flower/light constellations Twins, Scales or Waterman.[171] Cut the nettles close to the ground with a scythe early in the morning and let them lie until the afternoon so that they wilt slightly, as Steiner directed.[172] Slightly wilted nettles will attract cattle and other grass-eating animals, especially if the nettles have been cut on working pasture. Scything stinging nettles from the same spot repeatedly over a number of years is foolish because the plants will eventually die off, not having been able to develop sufficient foliage for assimilation. Leave alone areas in and around the densest nettle colonies which may contain birds' nests.[173]

Steiner said the freshly wilted stinging nettles can be compressed a bit and then put straight into a pit in the ground, although a thin layer of loose peat or something similar around the sides helps separate the nettles from the surrounding soil. 'Bury them right in the ground, but mark the place carefully so you do not just dig out plain soil when you come back for them,' he said.[174] The four sides and even the base of the pit can be lined with a layer at least 5 centimetres thick of peat and, once filled, the dug-up soil can be placed on top.[175] Filling the pit is easier if the nettles are bunched tightly or are stuffed into a coarse sack (e.g. burlap) first, to separate the nettles from the peat layer, and this also makes scraping away any moss or peat used to line the pit when excavating the preparation much easier.[176]

Stinging nettles are quite high in both protein and iron and so attract earthworms. An unglazed earthenware pot or clay pipe used in field drainage makes a stronger barrier than sacking.[177] You can first place a piece of wood cut to fit the diameter of the tile at the bottom of the upright tile. This wood plug maintains space for a clay seal inserted later. Then peat moss can be dropped in and tamped with a wooden tamper to form a plug 2 to 5 centimetres thick. The stinging nettles are then added, and tamped periodically as before. When the pipe is nearly full, the second circle of screening is placed on top, followed by more moss, leaving a 2-centimetre space at the top which is filled with moist clay. The wooden plug at the bottom is then removed and the space left there is also filled in with clay. At least 30 centimetres of excavated soil should cover the buried stinging nettles in the pit.

Stinging nettle's 'iron radiation'

Steiner said the nettles should spend 'winter and also the following summer in the ground; they need to be buried for a whole year. Then you will have a substantiality that is extremely effective.'[178] The implication here is that harvesting stinging nettles in late spring, burying them immediately and digging them up exactly a year later risks making an immature, less fully transformed and thus less powerful preparation because when Steiner said the stinging nettles must spend 'the following summer underground' he meant the whole of the second summer, up to and until autumn equinox.[179]

Steiner said that stinging nettle was a real jack of all trades, capable of doing many different things.[180] First, stinging nettle plays an important role in assimilating and incorporating the spiritual because it contains sulphur, like chamomile 503 and yarrow 504, and so reinforces those two preparations.[181] Carrying the spiritual into soil and food crops for humans is the fundamental goal of biodynamics. Stinging nettle also carries the radiations and currents of potash and calcium, Steiner-speak for stinging nettle's ability to attract, draw in and hold elements, but it also has 'a kind of iron radiation' that works in nature in a very similar and beneficial way to the iron radiations blood carries in and around the heart in the human organism. Steiner attributed stinging nettle's iron radiation to the plant's 'marvelous inner structure', meaning its ability to focus its energies not on its roots or flowers but on its leaves which are this plant's central organ or heart, and which Joly's 'sensitive diplomat' comments above specifically referred to.

Iron radiations are what keep stinging nettle's own sulphurous protein forces in check, as well as those brought to the compost by both the yarrow 502 and chamomile 503 compost preparations. Iron prevents sulphurous protein in our blood becoming toxic to us. In a similar way the nettles' iron process creates quality protein by organizing the chaotic forces of sulphur.[182] Steiner pointed out that sulphur is 'the element in protein that plays the role of mediator between the physical in the world and the omnipresent spirit with its formative power.'[183] For sulphur in stinging nettle to be beneficial it must be balanced by iron – note here we are talking about iron processes rather than iron itself which is present in stinging nettle only in limited amounts[184] – and that only happens if the nettles spend the winter in the earth and also remain there throughout the entirety of the following summer. This is when the August meteor showers or Perseids (originating from the constellation Bull) send meteoric iron forces across the northern hemisphere, reining in the sulphur forces of summer there.[185] Producers in the southern hemisphere may choose to leave the stinging nettle preparation buried throughout the Quandrantids meteor showers which occur there in January. Exposing the buried stinging nettles first to the forces of winter, when the earth is most inwardly alive, allows them subsequently to be able to absorb and retain the meteoric iron forces present during the whole of the following summer.

Steiner said that adding the stinging nettle 504 preparation to compost will make it 'more inwardly sensitive and receptive, so that it acts as if it were intelligent and does not allow decomposition to take place in the wrong way or let nitrogen escape or anything like that. This addition not only makes the manure [compost] intelligent, it also makes the soil more intelligent, so that it individualizes itself and conforms to the particular plants that you grow in it. Adding *Urtica dioica* in this form really is like an infusion of intelligence for the soil.'[186]

Rudolf Steiner said the stinging nettle 504 preparation was like 'an infusion of intelligence for the soil'.

Storing the stinging nettle 504 compost preparation

The marked decomposition the stinging nettles undergo during burial means that even a large pit fully filled with compacted nettles will yield only a relatively thin layer of finished preparation at the bottom of the pit when it is opened.[187] This thin layer resembles a black compost which should be 'beautifully colloidal'.[188] Courtney says 'the blacker the end product is, the higher the quality it is likely to have ... if one takes a small pinch of the finished preparation between one's thumb and finger and rubs vigorously, one will get a greater or lesser degree of smearing. The more the pinch of stinging nettle disappears or disintegrates into a stain or smear between thumb and finger, the higher the quality it has in all probability.'[189] Proctor suggests breaking up any untransformed stalks by putting the preparation through a sieve before storing it. 'There should be no problem with it drying out during the settling-down stage', he says.

Using the stinging nettle 504 compost preparation

This preparation can be produced in large quantities and can be used generously.[190]

Oak bark 505

Oak bark is perhaps the most difficult biodynamic compost preparation to get one's head around (a weak pun whose meaning will become clear by reading on). It is also the most difficult to make. Inserting the bark into its animal sense organ sheath is fiddly; closing the sheath even more so since this preparation must be buried in a place involving the dynamic of running water. To complicate matters even further Rudolf Steiner never fully explained why he stipulated this preparation's sheath must be an animal skull.

The oak bark preparation[191] cannot, however, be skipped over because it 'provides what is necessary for plants to be upright, and well formed and it develops the farm's immune system, conferring disease resistance on crops'.[192] The oak bark preparation's role in making individual plants grow in the ideal way helps the entire farm organism or vineyard function properly as a whole. By taking the place of the brain in the skull within which it is buried the oak bark can be said to become the 'brain' of the compost/farm organism. From an anthroposophical perspective (see Chapter 1) this means the oak bark preparation brings to compost what crops grown in biodynamically composted soil do to humans, namely a consciousness, perception and sensibility which breeds physical, emotional and spiritual wellbeing.[193] The oak bark preparation 'firmly fixes the egoic force, the individuality of the farm, giving it inner strength, but also outward form', says Lovel.[194]

Steiner's contention was that soil with the right level of calcium (force) helped plants stay healthy and disease-free. Oak bark's rich calcium content makes it ideal preparation material. Steiner said that adding the oak bark preparation to compost 'will truly provide the forces to prevent or arrest harmful plant diseases'.[195] As Proctor explains, 'the calcium brings harmony to the form of the plant and does not allow excesses of rampant growth to develop. This condition often arises when the moon forces are working too strongly and after a wet period. For the calcium to have a healing effect, according to Steiner, it should be in a living form [meaning as a biodynamic preparation in compost], and it is of no use if applied to the soil as a mineral.'[196]

So while Steiner could have chosen another calcium-rich plant like seaweed, for example, for this preparation he insisted on oak bark because it 'represents a kind of intermediate product between the plant and the living earth element'.[197] This recalls Steiner's idea that tree trunks are like soil raised up above the earth's surface, tree bark is like the soil surface, and tree branches are like annual plants growing out of this soil (see also Chapter 5, winter tree paste). Steiner's reasoning was that 'of the many forms in which calcium can appear, the calcium structure of oak bark is the most ideal. Calcium … creates order when the etheric body [the plant's own inner growth force] is working too strongly, so that the astrality [which balances the etheric] cannot influence whatever organic entity [meaning the crops] is involved. Calcium in any form will kill off or dampen the etheric body [by providing this balancing astrality] … but when we want a rampant etheric development [that which encourages an excess of disease-causing bacteria and fungi, or weeds] to contract in a beautiful and regular manner, without any shocks [like hormonal stress], then we need to use calcium in the particular form in which it is found in oak bark.'[198] The overall message is that plants only get sick if the realm they grow in, the earth, has been tampered with.[199] Putting oak bark preparation 505 in fresh compost and then spreading the matured compost on the soil would engender this correction, first in the soil, then in the plants growing there and finally in us when we eat the crops.

As noted above, Steiner never said why the skull is used in this preparation, but Proctor says that 'a connection to the skull as a moon vessel can be seen if one studies embryology. In all embryo development, which Steiner points out is a moon-growth process, the skull appears first. Another example of the moon-growth process is seed germination, which is enhanced at the time of the full moon ... As we are trying to regulate the calcium [to make it work in the right way when added to compost], we surround the bark in a bony "container" [the skull] which is built up from calcium. Placing the skull and the bark in a watery treatment will bring a balancing effect between the [potentially excessive] moon forces and the calcium [reining them in].'[200]

Placing the oak bark preparation in a watery element is what connects it to the forces of the outer planets: Mars, Jupiter and Saturn. These planets support or enhance the silica processes: supporting nutritive value, taste (ripeness), colour, aroma and so on in crops. This balances the inner planets between the earth and the sun – namely Mercury, Venus and the moon – which support or enhance the calcium processes of reproduction and growth. As Lovel says, 'it might seem intuitively obvious that with its high calcium content [the oak bark preparation] establishes the skeletal framework for the farm organism.'[201]

To summarize: plant material with a unique quality and concentration of calcium, the bark from a highly evolved plant, the oak, is being surrounded by mud (the earth) and water (the moon) within a calcium-rich animal part, the brain cavity, of highly evolved animals such as cows to produce a preparation which will strengthen forces in the earth which counteract tendencies in plants to become diseased.[202]

The significance of oak bark

Of all the biodynamic compost preparations, oak bark is unusual because bark is used rather than flowers (yarrow, chamomile, dandelion, valerian) or vegetation at or close to flowering (stinging nettle). However, Courtney[203] says bark can be seen as the manifestation of the force or nature of a flower. It is at the flowering stage that the astral most closely approaches the plant, as flowers send messages via the air by way of scent. Steiner's aim with oak bark was to provide a preparation capable of correcting any imbalance in the etheric by stimulating a strong astral energy. Within the nerves–senses organism of the human being, the astral dominates as well. It is within that nerves–sense pole that Steiner speaks of the need for a state of nearly death-like existence for the nerve cells, in order to bring about the possibility of human thinking (brain cells die when we think and have no power of regeneration).[204] In order to create an 'organ of thinking', as it were, for the compost pile – and ultimately for the earth treated with that compost – Steiner suggests a substance so closely approaching a state of deadness that it can be seen as the nearest still-living substance within the plant kingdom which is akin to the brain cells. This substance, the outermost layer or rind of bark from the oak tree, is then placed within the skull whose membrane lining is called the *meninges*. The subsequent burial in a watery pit during the wintery months when the earth is most inwardly alive serves to create a brain-like organ for first the compost, and then the soil because – as von Wistinghausen *et al.* point out[205] – the bark, protected by the calcified cranium, goes through a rotting process. Just as the brain is surrounded by the fluid in which it is floating, a watery earthy fluid also surrounds the oak bark within the bark-filled *meninges* when it is inside the animal skull. It is only this step of creating an organ of thinking, or consciousness, that allows the possibility for a farm organism to develop and to become an individuality. It is only when a farm individuality or a farm ego is realized that farmers achieve the purpose of the biodynamic preparations, namely that the earth and ourselves may be healed.

Making the oak bark 505 compost preparation

The oak tree specified by Rudolf Steiner was the pedunculate (English) oak or *Quercus robur*.[206] However, as this only grows successfully in temperate climes, substitutes can be used. For the North American continent, Ehrenfried Pfeiffer suggested the white oak, *Q. alba*, as being most like the English oak.[207] Both the sessile (*Q. petraea*) and red (*Q. rubra*) oaks are also suitable.[208]

Only the outer bark should be collected. Avoid bark from the side of trees facing away from the equator as they are likely to be covered with moss. The pithy material immediately beneath the bark, the cambium layer, should not be used because this part of the tree is still alive.[209] The bark 'acts as an intermediary between the dead and the living'[210] and is what the tree rejects 'toward the [lifeless] mineral world while remaining at a stage that is higher and more alive.'[211] Hence it provides calcium in the ideal 'living' form Steiner wanted. The oak bark preparation will, over a period of time, raise the pH of the soil without the addition of lime, says Proctor,[212] adding that agricultural lime applied to the land is made more effective where the oak bark preparation is used, and consequently less lime is required.

The oak bark is collected when its calcium content peaks, from late summer through early autumn, and from living trees at least thirty years old with trunks at least 30 to 50 centimetres in diameter.[213] The bark should be collected in the afternoon.[214] Bouchet advises collecting it under a descending moon and when the sidereal moon stands in a root/earth constellation: Virgin in the northern hemisphere and either Goat or Bull in the southern hemisphere.[215] The bark can be scraped off using a metal beekeeper's hive tool or a short-handled hoe, and allowed to fall into either a cloth left on the ground around the base of the tree or into a cardboard box which can be shaped to the contours of the tree.

Rudolf Steiner indicated that the skull of any domesticated – meaning farm – animals, such as bovines, sheep, goats, pigs or horses would do for the sheath.[216] Skulls from animals killed on the farm or from neighbouring biodynamic farms are ideal.[217] The cranial cavity of the sheep is the biggest in terms of volume. The skulls should be fresh or at least freshly frozen because, says Courtney, 'the most important single factor in using any skull is to ensure that … the membrane lining the skull cavity, the *meninges*, is intact. In a certain sense, the true sheath for the oak bark is not the skull, but rather the *meninges* … to remain within the realm of the living as Steiner encourages us to do can only be accomplished for this preparation by preserving … the *meninges* membrane.'[218] Courtney also says that leaving the skull's snout and jawbone intact gives a better result, and that as the slaughterhouse procedure for pigs involves sawing the skull and splitting the brain cavity into two separate pieces this means the skull must subsequently be wired back together. Skulls from kosher-killed animals have no bullet hole so the *meninges* lining will be intact.

Cleaning the brain cavity of flesh and brain material reduces the risk of polluting watercourses should the skull be buried in one. The *meninges* lining remains intact if the skull is gently cleaned using water, perhaps with a thin copper pipe soldered to the end of a water hose with

Tree huggers feel at home collecting the oak bark 505 preparation

a shut-off valve.[219] Cleaning the skull by putting it in a compost pile for microbes to strip it of flesh, or in a tree for birds to do the same, risks perforating the *meninges* lining or even completely destroying it.

However, Demeter International's current production standard stipulates the skull be cleaned before it is filled with oak bark by being 'placed in a closed container filled with sawdust and left for a period of time during which it is cleaned of any fleshy remains by means of a process of microbial maceration. After the skull is removed waste material is disposed of in accordance with current regulatory requirements.'[220]

The oak bark should be chopped or ground to a crumb-like consistency,[221] perhaps by passing it through window screening (8 mesh) or through an old cornmeal grinder in the case of small volumes, or a grain grinder for larger ones. It can then be moistened with oak bark decoction (see Chapter 6). The decoction should be allowed to cool so as not to scald the life force within the ground oak bark when it is moistened.[222] The ground bark is incredibly absorbent, indicating its capacity for reining in the excess water forces which can lead to excessive growth and thus diseases in plants.

When the ground oak bark is damp but not dripping it can be packed into the cranial cavity via the *foramen magnum*, the opening at the rear base of the skull, using a wooden dowel if necessary. Once filled, Steiner recommended the skull be closed with a piece of bone, from say the shin or jaw of the animal, giving an exact fit. The eyeball cavity, the mouth and any smaller openings can also be sealed in the same way. Using moistened clay instead of bone can be much easier, however. The ground oak bark should be packed tightly enough to leave no air pockets. The filled skull can be dropped in a water butt. If it sinks, the skull will be free of both brain tissue and air pockets and is ready for burial.

Steiner said to place the skull in a relatively shallow hole in the ground, covering it with loose peat.[223] Netting may be needed to prevent dogs or foxes from digging up the skulls. As much rainwater and snowmelt as possible should flow through the hole, perhaps by allowing a drainpipe or gutter from a roof to eject water over the hole;

Freshly excavated oak bark 505 skulls

or, as Steiner suggested, the filled skulls could be placed in a rain barrel where water can constantly flow in and out. Plant matter such as leaves and grass that will decay should be added to the barrel 'so that the oak bark in its bony container lies in this organic muck for the whole winter'.[224] If the skull is left in a swamp where there is a small flow of water it may wash away unless it is placed in a wire-netting cage attached to a post.[225] As burying skulls in ponds or the beds of streams may cause problems with water authorities, it is best to create a suitable place on your own land.

The oak bark preparation is buried in autumn, left for one winter, and lifted around spring equinox. Bouchet advises retrieving the filled skulls when Mercury stands in the fruit–seed/warmth constellation of Ram.[226] Courtney says to hose off the skulls before trying to open them to avoid contaminating the finished preparation with earth.[227] The finished preparation is then removed using a 'nit picker' or piece of wire that has a hook bent into one end of it. When as much of the preparation has been scraped out as possible the skull is then turned over and tapped vigorously so that the rest empties out via the *foramen magnum*. The animal skulls should be used only once. They can be disposed of by putting them in the centre of a compost pile shortly after it is constructed, where they will soon disintegrate completely.

Storing the oak bark 505 compost preparation

Proctor says that once retrieved, the oak bark preparation should be put in a glass jar (it will smell because it is still anaerobic) and turned every day until aerobic activity takes over and it smells quite sweet (up to fourteen days).[228] White mycelium growing on it are of no concern, he says.

Using the oak bark 505 compost preparation

The contents of a single skull should produce enough preparation to treat 300 hectares of cropland.[229] Courtney gauges one unit of the oak bark preparation as a rounded teaspoon by volume, which is about 2.5 to 3.0 grams by weight, enough for 10 to 15 tons of compost, he says.[230] A rounded tablespoon is a more accurate volumetric measure for those who prefer to insert less finely ground and thus larger flakes of raw oak bark into the animal skull.

California winegrower Jimmy Fetzer of Ceàgo del Lago holding the oak bark 505 preparation

Dandelion 506

When you next swipe a dandelion puffball with your non-standing foot while out jogging, the light snap of its latex-filled stem against your latex-filled running shoes being heard not felt, remember Rudolf Steiner said the little seed parachutes now gliding in your slipstream are potential messengers from the celestial sphere. Steiner saw in the dandelion a plant capable, if prepared according to his biodynamic instructions, of bringing a unique enabling quality to crops, giving them the power to sense exactly what they need from both their local and the wider environment, and even in fact to attract it.[231]

Steiner said that dandelion had this power, which he described as being able to 'draw in the cosmic factor', because of its perfect potassium–silica relationship. This enables it to impart to the compost the capacity to draw in those forces which are released through silica, specifically those exerted by the outer, silica-supporting planets Mars, Saturn and Jupiter, with the latter especially symbolic of something abundant and prolific, but not excessive. If dandelion was a plant dominated by earthly forces its root system would sprawl along horizontally in the topsoil. Instead dandelion's long, carrot-shaped, latex-filled tap roots punch holes in compact ground (your driveway, perhaps?) to open it up again both to visible light and invisible forces streaming in from the celestial sphere.

The upper part of the plant consists of bright yellow flowers which follow the sun like a sun clock. After flowering, the silica in dandelion is expressed when its seed-bearing pappae are borne on the wind. We blow these spindly, light-as-a-feather parachutes into each other's faces as children, little knowing their translucent appearance results from an abundance of 'fine siliceous cellular tissues', according to von Wistinghausen *et al.* who add that 'the silica coming to expression in the dandelion … makes this plant a special sense organ for the light. Silica supports sensory functions in human and animal skin and sense organs, above all in the eyes. In the plant, and above all in the dandelion, this type of sensory function arises from interaction between silica and potassium.'[232]

In dandelion, silica's role in attracting cosmic – perhaps that's better put as brightening – forces into the soil is enhanced by sheathing its flowers in the mesentery/peritoneum of a cow. Whereas the animal's digestive tract is open at either end via the mouth and anus which expose it to earthly forces, the mesentery/peritoneum holds in the animal's digestive organs but is completely sealed off from earthly influence. As a result the digestive forces released by the cow but which are retained within her by the action of her horns and hooves (see horn manure 500, above) means celestial influences can be especially strongly focused on the dandelion when this is enclosed in the mesentery/peritoneum and ultimately buried. Once dug up, Steiner said the dandelion will be 'thoroughly saturated with cosmic influence'.[233] Adding the dandelion preparation to the compost pile is a way of releasing light into the soil once the compost is spread there, meaning light in the sense of allowing the plant crops to 'see' by sense, not as if they had been given a torch.

Making the dandelion 506 compost preparation

Proctor[234] suggests the dandelion flowers (*Taraxacum officinale*)[235] are picked in spring, early in the morning before the sun is too high and before the centre of each flower has opened too fully. The flowers should be about half open, with the middle petals still folded inward like a kind of crown, or small cone. Flowers which are too open when picked will become very fluffy and white when dry, as if fully matured and running to seed. They should be discarded. Dry the freshly picked flowers by leaving the filled picking containers outside in the sun so the flowers wilt before being moved into airy shade and being spread in a single layer on a drying frame or on paper,[236] or they can be moved indoors directly and be turned in the early drying stage until quite dry. Once dry they can be kept in brown paper bags or vacuum packs until it is time to use them to fill the animal sheath.

The animal sheath for this compost preparation is something of a conundrum because when Steiner discussed it in his 1924 *Agriculture* course he directed that the bovine mesentery be used. However, in a subsequent question and answer session he said, 'to my knowledge, the mesentery means the peritoneum'.[237] Malcolm Gardner's notes to his co-translation of Steiner's 1924 *Agriculture* lectures state that the abdominal cavity of mammals is lined with a membrane called the peritoneum.[238] The stomach and intestines are suspended from the dorsal side of the abdomen by a sinuous fold in the peritoneum known as the dorsal mesentery (the ruffle, to butchers). The portion of the mesentery supporting the stomach (rumen) is itself folded and extended to form an apron over the front of the abdomen; this apron is known as the greater omentum (the caul or net, to butchers). It is not clear whether Steiner had a particular portion of the mesentery in mind, Gardner concludes.

Christian and Eckard von Wistinghausen, whose father Almar was the youngest person to attend Steiner's 1924 *Agriculture* course, describe one

A fresh bovine mesentery

possible approach.[239] 'The animal organ used to contain the dandelion flowers is the bovine mesentery or caul fat (great omentum). The natural position of stomach and intestines within the coelum is complex and normally inaccessible. The omentum is the first thing a skilled butcher removes, while the animal is still lying on the ground. It is the great fold of the peritoneum, starting at the rumen and enclosing the whole intestine like an apron on the abdominal side. It lies loose except where it is joined to the stomach. When the carcass has been raised up, the butcher first of all removes the stomachs and then the whole intestine. As it lies spread out one sees the mesentery of the small intestine along the end of the peritoneal fold. If the small intestine is cut away, the peritoneum forms coils corresponding to those of the intestine. The ruminant colon is a spiral within this fan-like peritoneal fold. It has to be cut away from one side before the mesentery can be used for our purposes. The omentum may also be used. The peritoneum is the vehicle for the nervous system throughout the metabolic region of the animals. The dandelion flowers are wrapped in the lean middle part of the mesentery … moisten the dried flowers with a tea or juice expressed from whole dandelion plants minus the roots, using just enough to have them slightly damp, like fresh flowers that have wilted a great deal. Cut pieces of mesentery or omentum into squares, 20 to 35 centimetres in width. Put 2 to 5 handfuls of dandelion flowers on each square, pressing them together, and folding the corners of the material up around the flowers so that they are completely enclosed.'

The dandelion-filled pouches or pillows are usually buried in autumn. In handwritten notes for his 1924 *Agriculture* course Steiner suggested that the dandelion compost preparation would benefit from spending six months hanging in a tree before being buried in the ground, in a similar way to the stag's bladder for the yarrow 502 preparation and the cow intestine for the chamomile 503 preparation.[240] The dandelion would thus be exposed to the sun for at least part of the summer whilst also benefiting from the crystallizing forces present in the earth over winter (see horn manure 500, above). In this case dandelion flowers picked in summer would dry over winter for use the following spring. Once filled, the sheaths are buried during a descending moon,[241] and if possible also following the sidereal moon when this stands in a root/ earth constellation: Virgin in the northern hemisphere and either Goat or Bull in the southern hemisphere.[242] Protect burial pits from burrowing animals in the same way as described for the yarrow 502 preparation, above.

Steiner indicated that the dandelion preparation should be dug up at spring equinox, at the same time as you would dig up the horn manure 500, although you may wait another six weeks or so.[243] For the northern hemisphere Bouchet advises digging up the preparation in April following its burial,[244] when the planet Mercury stands in the fruit–seed/warmth constellation of Ram.[245] Proctor says the mesentery should still be intact and not broken down.[246] He advises cutting it open and removing the bulk of the dandelion preparation before scraping any broken-down flowers off the folds of the skin. Courtney says that a more humus-like preparation is obtained if the peritoneum, rather than the mesentery, is used as the sheath, implying that the structure of the blossoms is still visible and thus less humus-like when the mesentery is used.[247] François Bouchet says that the dandelions should retain both a recognizably identifiable form and smell but that the smell will have become 'much more animal'. The sheath that held the dandelion flowers can be crumbled or chopped up and added to a liquid manure.[248]

Filled mesentery pillows for the dandelion 506 preparation waiting for hanging, then burial

Storing the dandelion 506 compost preparation

Proctor advises storing the dandelion compost preparation in a glass jar and checking it regularly, if necessary gently aerating it for the first few weeks after lifting to prevent it becoming anaerobic.[249] The jar lid can even be left off for the first few weeks.

Using the dandelion 506 compost preparation

Whenever I make a hole in the side of the compost pile into which the dandelion preparation is to be added, I feel like a plumber reaching into a dark recess in which a blockage needs attending to. This, ultimately, is dandelion's role: regulating, balancing and harmonizing, helping crops to draw their nourishment from their broader environment and overcome obstructions to the flows of forces in the farm's surroundings.[250] A 30 centimetre by 30 centimetre dandelion-filled peritoneum/mesentery pillow should produce enough of this compost preparation to treat 100 hectares of cropland.[251]

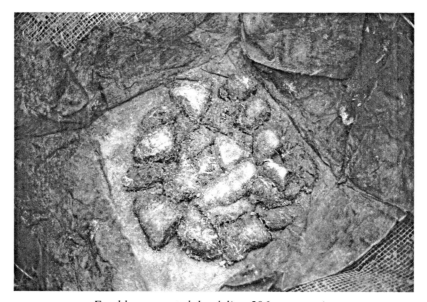

Freshly excavated dandelion 506 mesenteries

Valerian 507

There's nothing more satisfying at the end of an autumn day's composting than dipping your cold hands in a container of warm water and stirring in a few drops of extract of valerian flowers, backwards and forwards. The resulting pale amber liquid is then sprayed over the finished compost pile. Like icing on a cake,[252] the valerian seals in the beneficial forces carried by the other five compost preparations so that when the compost is spread these forces can be released to the soil.

Making the valerian 507 compost preparation

Valerian (*Valeriana officinalis*) is a perennial herb (Valerianaceae family) native to Eurasia which flowers in early summer.[253] Its tiny white flowers are intensely perfumed, appear in clusters and open individually over a period of time. About two weeks after the first flowers have opened the first few petals drop off, indicating that seed formation has begun and flower picking should also begin. At this stage about half of the blossoms on a given flower head should have opened.[254]

Biodynamic literature generally suggests picking the flowers in the early morning. However Bert de Liefde argues that after 5 p.m. is optimum, citing Dr Ehrenfried Pfeiffer's sensitive crystallization tests (see p.104) on juice from valerian flowers picked at different times of the day.[255] De Liefde concludes that 'picking flowers in the morning [will] yield more juice, but the juice does not have the full, rich scent obtained from evening picked flowers'. Picking the blossoms near summer solstice and preferably when an ascending moon (see Chapter 7) stands in an air/light–flower constellation (Waterman in the northern hemisphere, Scales in the southern hemisphere, and Twins in either hemisphere)[256] is said to enhance greatly the keeping quality of the preparation.[257] Ideally leave some flowers for insects.[258]

Common practice is for the flowers to be chopped or finely ground using a pestle and mortar, perhaps by placing the pulp in the toe end of a nylon stocking, twisting it up tightly and squeezing the package in a vice to extract a green or coffee-coloured juice. Or, if flowers are in short supply, they can be wetted with a little rainwater or snowmelt rather than well water[259] (meaning water previously exposed to the sun), and left in a glass jar on a sunny windowsill for up to a week before the mixture is pressed or strained through muslin[260] into sterilized 20- to 50-millilitre bottles.[261]

De Liefde argues that making the valerian extract in either of these ways results in a poor yield and a preparation whose concentration is unknown. His alternative method is to process the flowers the same evening they are picked, snipping off all stem parts but leaving the green calyx in place. 'This yields a batch of small clusters of flowers and buds,' he says. Using a funnel and stick, these are stuffed into glass bottles which can be sealed with a cork: a 150-millilitre bottle for up to 30 grams of flowers, a 300-millilitre bottle for 30 to 60 grams of flowers and a 600-millilitre bottle for 60 to 120 grams of flowers. Once the flowers are inserted, add distilled water until the bottle is full, but leave an air space below the cork; boiled but cooled drinking water will do in the absence of distilled water. Make notes of the weights and quantities. Tie the cork down with twine to prevent it from popping out. Then hang the bottle that same evening in a sunny position in a tree and leave it there for three days. 'A yellow amber fluid which is quite clear is visible below the flowers, which are now slightly discoloured. Occasionally the fluid is green and turbid … the resulting preparation is of lesser quality, though still usable,' he says.[262] After three days in the tree the bottle's contents are drained into a fresh container via a filter like a piece of fine terylene fabric. Extract any juice contained in the solids by pressing and combine this with the larger volume of free-run juice.

Stirring valerian 507 before applying it to compost

Storing the valerian 507 compost preparation

Those who make their valerian extract using either a pestle and mortar or by using water to extract juice from the pulp store it directly in bottles which either have the caps not quite screwed on, or in bottles sealed with a fermenting cap so air cannot get in but any gases released by an acid fermentation in the juice over a period of about six weeks can escape.[263]

Using De Liefde's alternative method you would first store the freshly expressed liquid concentrate in bulk (i.e. in 200-millilitre bottles in an insulated box) rather than in individual small vials. 'A spontaneous fermentation usually occurs in the raw preparation,' he says. This can be tempered by temporarily storing the liquid in squat rather than upright vessels, keeping these in the refrigerator at 7 to 10°C until the fermentation has died down, and straining out any impurities or plant tissue sediments on a regular (even daily) basis. You can use filter paper circles and a funnel, but speed is of the essence because exposure of the liquid to sunlight or air will have a negative effect on the finished preparation. To speed drainage, the filter paper can be formed into the cone shape of the funnel using galavanized bird netting placed between funnel and paper. The funnel can be covered, too, to exclude air. De Liefde says, 'After perhaps three weeks of daily inspection and occasional straining, the fluid will become quiescent and remain transparent and clear. That is soon after midsummer. The fluid can be left in the refrigerator for a few more weeks if desired, before it is bottled. Store in 200 ml bottles … Leave space for an air bubble in each bottle and drive the [tapered] cork right into the neck as in a wine bottle …[the smell] should be rich as of the fresh flowers, but with an additional tartness which has arisen during processing.' The corks should be kept wet and thus more airtight by lying the bottles on their sides. Once the bottles are opened so that concentrate may be drawn off for compost or spraying directly on the land or crops (see Chapter 6, valerian tea), any remaining unused liquid concentrate should be decanted into vials and topped up with distilled water. The valerian extract matures with storage and may be kept for several years and until it starts losing its scent. Ideally the juice is only drawn off its sediment on the day it is to be used.[264]

Using the valerian 507 compost preparation

The standard practice is to dilute the valerian liquid concentrate in water to a 5 per cent solution. One 10- to 15-millilitre portion, which is enough for 10 to 15 tonnes of compost, is dynamized (see Chapter 4) into 10 to 15 litres of warm (37°C) rainwater for 10 to 15 minutes.[265] De Liefde, however, says that you can only know precisely how much valerian you are using by weighing the flowers and measuring the fluid quantities when making it, so as to calculate the concentration of the preparation. The flowers must be free of dew or rain at the time of weighing.

De Liefde says the starting weight of the fresh flowers placed in the bottle (*a*) minus the weight of pressed-out cake (*b*) should be divided by the starting the weight of the fresh flowers (*a*) and that the figure for (*a* minus *b*) divided by (*a*) should be 0.5 to 0.7. The next figure to obtain (expressed in millilitres) is the pressed out fluid increase (*c* minus *d*), where (*c*) is the total quantity (in millilitres) of fluid collected in the preserving jar and (*d*) is the quantity (millilitres) of distilled water added to the flowers. De Liefde notes that (*a*) minus (*b*) is always more than (*c*) minus (*d*) for two reasons: the specific gravity of the juice is greater than that of water (which has a specific gravity of 1) and a small quantity of fluid is retained in the fabrics and vessels used.

The concentration of the preparation in the preserving jar (expressed as grams per millilitre) is (*a* minus *b*) divided by (*c*). The usual range for (*a*) minus (*b*) divided by (*c*) is 0.05 to 0.2 grams per millilitre. De Liefde concludes by saying that the standard dose or portion of preparation (which would be sufficient for 10 to 15 tonnes of compost) is 0.5 grams of neat juice equivalent. This means that *c* (millilitres) divided by 2 (*a* minus *b*) has to be placed in each glass vial, which is then topped up with distilled water and sealed with a white plastic snap-on top.

Once the correct quantity of valerian liquid concentrate has been stirred in water it can be sprayed over the compost pile using a back-sprayer or flicked over it using fingers or a small brush. Some biodynamic growers spray half the recommended volume over the pile and pour the rest into a hole made in the compost heap, in the same way as for the five solid compost preparations 502–506.

Most biodynamic growers spray the valerian compost preparation on newly made compost piles, before the piles are covered with straw for winter. However, it is worth considering the argument that the prepared (meaning dynamized/stirred) valerian should in fact be sprayed only several months later and when the pile of raw uncomposted material has actually become transformed into finished humus-rich compost.[266] Rudolf Steiner said compost treated with the valerian preparation would activate the phosphorus forces of light. Plants need phosphorus to attract the sunlight their leaves need for photosynthesis. Adding valerian-enriched compost to soil stimulates the phosphate process there and mobilizes the phosphate-activating bacteria.[267] In effect, the valerian brings light forces into the dark underground of the soil, like the bright white light of a burning phosphorus match.

Of the six biodynamic compost preparations, valerian is unique in two ways. First, it is the only one not to be buried underground – a process which concentrates forces within a substance. Second, it is the only one not to be used in solid form but sprayed as a liquid – a process which allows forces contained within a substance to be released. So if you want to release those forces brought to and preserved within the raw compostable material by the five other preparations as it composts, then spray valerian on finished rather than newly made piles; or simply hedge your bets and spray it twice, before and after. Valerian is the phosphorus 'light' that switches the compost on.[268] It makes the metaphorical blood within the compost and farm organism – and perhaps the real stuff in your hands on a cold day – finally flow.

THE BIODYNAMIC PREPARATIONS 500–508: A SUMMATION[269]

Rudolf Steiner saw plants as organisms which were connected not only to every surrounding plant but also to celestial bodies in the far reaches of our universe. The biodynamic preparations 500–508 stimulate the growth and reproductive forces plants need to realize their full potential to connect to what surrounds them on both a physical level – the three realms of nature, namely animal, mineral and vegetal – and on a forces level. Growth forces which help vines' self-expression below ground are charged by horn manure 500 whose lime/calcium forces work with the earth and water elements promoting *terroir*-driven wines. Balancing these growth forces above ground are the hardening, ripening forces of silica. These silica forces work with air and warmth to promote both crop taste and crop quality via horn silica 501 and crop health via common horsetail 508 which keeps fungal diseases at bay. Combined, these three field spray preparations collectively promote *terroir*-driven, ripe-tasting, healthful wines.

Steiner said that the elements of carbon (earth), oxygen (water), nitrogen (air), and hydrogen (fire) bind together to make food and wine with the help of sulphur and phosphorus, which bear light forces. Potassium, magnesium and the trace elements are also realized as forces and not just substances, and the processes of these forces are what Steiner said biodynamic farmers should be focused on.

The six compost preparations 502–507 allow forces to permeate the vineyard which promote the mobilization of trace elements via sulphur and potassium forces (yarrow 502); individual plant health via calcium forces (chamomile 503); overall soil health via iron and silica forces (stinging nettle); overall farm health via calcium force (oak bark 505); the sensitizing and balancing of plants to their precise surroundings, namely the plot of earth they grow in as well as exactly where that plot of earth relates to in the greater celestial sphere via silica forces (dandelion 506); the preservation, enhancement and release of forces brought by all the other preparations via phosphorus (valerian 507). Biodynamic preparations should enrich both the vineyard and the winegrower in varied, balanced, life-enhancing ways to promote *terroir*-driven wines which people should find taste ripe and clear but which are also complex, original and vital.

Table 2: The six biodynamic compost preparations 502–507

Preparation name	Number	Made from	Animal sheath	How and when made	State	Role	How used	Volume of prep/10 metres3 of compost	Volume of preparation/ha*
Yarrow	502	Yarrow flowers (Achillea millefolium)	Bladder of red deer stag (Cervus elaphus)	Expose filled bladders to sun from spring to autumn, then bury until late spring/midsummer the following year	Solid	Sensitizes vineyard to celestial cycles, natural forces	Insert in solid form into compost pile	1-2 cm³⁺	Contents of one stag's bladder per 250 hectares of cropland
Chamomile	503	Chamomile flowers (Matricaria chamomilla)	Cow intestine	Expose filled intestines to sun from spring to autumn, then bury until spring equinox/early summer the following year	Solid	Allows retention and flow of living forces in vineyard	Insert in solid form into compost pile	1-2 cm³⁺	Contents of one 30cm intestine per 100 hectares of cropland
Stinging nettle	504	Stinging nettle leaves, shoots (Urtica dioica)	None	Bury freshly dried nettles at summer solstice, dig up in autumn the following year	Solid	Brings intelligence to soil, vines	Insert in solid form into compost pile	1-2 cm³⁺	No set rate. Use generously
Oak bark	505	Outer bark of pedunculate (English) oak (Quercus robur)	Meninges-lined skull of any domestic (farm) animal	Place stuffed skulls in watery environment in autumn, retrieve at spring equinox the following year	Solid	Regulates vine vigour	Insert in solid form into compost pile	1-2 cm³⁺	Contents of one skull per 300 hectares of cropland
Dandelion	506	Dandelion flowers (Taraxacum officinale)	Bovine mesentery / peritoneum	Expose filled mesenteries to sun from spring to autumn, then bury until spring equinox/midsummer the following year	Solid	Brightens soil, sensitizing vines to their physical surroundings	Insert in solid form into compost pile	1-2 cm³⁺	Contents of one 30 x 30-centimetre mesentery per 300 hectares of cropland
Valerian	507	Valerian flowers (Valeriana officinalis)	None	Make liquid concentrate from valerian flowers picked in early summer	Liquid	Aids retention and release of natural forces	Dilute concentrate in water to 5 per cent solution, stir for 10 to 20 minutes, spray on and/or pour into compost pile	0.5g concentrate	No set rate

* Source: Estimated quantities Demeter production standards for the use of Demeter Biodynamic® and related trademarks June 2015 (Demeter-International e.V.), p.44

3

BIODYNAMIC COMPOST 502–507

We tend to think that healthy soils are what make healthy plants. In fact the opposite is true.[1] Plants grow by being able to capture solar energy, and when they die and decompose they condense the intangible matter they took from the sky into those tangible substances which create both soil and soil's core constituent, humus. Composting is simply a way of getting microbes already present in disparate organic waste materials like plant leaves, plant prunings, plant residues (grape pomace), plants which have been eaten and digested by animals (manure) and other biodegradable waste first to decompose and then to break down into something more homogeneous, and finally to reconstitute what has broken down into something capable of feeding and enhancing soil in the ideal way. Compost improves soil structure, bringing both nutritional balance and beneficial microflora. Steiner said that 'in compost we have a means of kindling the life within the earth itself'.[2]

COMPOST'S ROLE IN THE SELF-SUSTAINING FARM ORGANISM

A fundamental tenet of biodynamics is each farm or vineyard should try to become a self-sustaining organism. Steiner said that 'a farm comes closest to its own essence when it can be conceived of as a kind of independent individuality, a self-contained entity … every farm ought to aspire to this state of being a self-contained individuality. This state cannot be achieved completely, but it needs to be approached. This means that within our farms, we should attempt to have everything we need for agricultural production, including, of course, the appropriate amount of livestock.'[3]

Lovel points out why Steiner attached such significance to farms having a balance of both animals and crop plants: 'Plants are formative. They build. Animals are transformative and they recycle or break down and rebuild ... the conventional investigator may say the animal breaks down plant tissues with digestive enzymes. No doubt he will acknowledge that in most cases there is a certain amount of chewing involved. But the BD [biodynamic] practitioner can only shake his head since he knows animal digestion and nervous development run parallel and both these activities are transformative. For one thing, how does the animal select what it chooses to eat?'[4]

As biodynamics concerns forces as well as substances, animal manure is seen to provide both substances, via enzymes and other organisms needed in compost (fungi, bacteria, and ultimately the preconditions for worms), and forces because of the nervous development the animal exerted when digesting plants to produce the manure in the first place (see also horn manure 500 in Chapter 2).

'Without the digestive activities of animals, plant fibres would be much slower to break down and return to the soil,' says Lovel.[5] Thus a diversity of both plant and animal species is needed to assure rapid recycling of plant materials in order for farms and vineyards to become successful self-sustaining organisms capable of transforming one-way waste streams born of an 'acquire, use, discard' mentality into a regenerative cycle of fertility.

Composting allows the recycling back onto the land of the by-products of wine production when micro-organisms (mainly bacteria) in manure are allowed to break down organic matter into gas (carbon dioxide), water and, most importantly, solid and stable humus. The two most obviously compostable by-products of winegrowing are cut vine prunings in late winter/early spring, and anything left over from winemaking every autumn, mainly the grape pomace, pressings, stems and lees. Other winery by-products which can also be composted include diatomaceous earth, bentonite finings, used filter pads and even shredded waste paper (a carbon source) from the office.

Grape prunings can, of course, be recycled by being left to decompose by neglect *in situ* where they fell in the vineyard. They decompose more quickly if first chopped or shredded into small pieces using a chipper mounted on the back of a tractor, and more healthily in biologically active soils stimulated by soil sprays like horn manure 500, Maria Thun's barrel compost 502–507 (or its derivatives) or stinging nettle liquid manure.

Grape pomace, however, is almost impossible to spread evenly across the vineyard because volumes of it are so relatively small (a trailer-load of grapes will produce a barrel of pressed pomace). In addition, pomace quickly turns vinegary, attracting fruit flies. Adding raw, uncomposted pomace and pressings to the soil is a far less efficient way of building the stable humus necessary for enhanced soil structure compared to proper composting, and risks not only failing to kill off, neutralize or deter undesirable pathogens, but also perhaps encouraging them.

Secondly, while all winegrowers see plant growth being shaped by tangible substances like soil nutrients (and even atmospheric carbon dioxide), only biodynamic growers also see the etheric formative forces described by Steiner as being equally fundamental. Vines become best sensitized to these only if the soils they grow in are treated with material composted in the presence of the six biodynamic compost preparations 502–507. While Maria Thun's barrel compost 502–507 spray or its derivatives, and Alex Podolinsky's prepared horn manure 500 + 502–507 spray (see Chapter 5) succeed in getting the six compost preparations onto the soil, neither are genuine substitutes for solid biodynamic compost 502–507, particularly in the case of vineyards recently converting from years of conventional winegrowing. They need restocking with organic matter. Biodynamic compost 502–507 is the best tool to enhance both the substance of the soil via more stable humus and the soil's capacity to hold and resonate the formative forces needed by the plants it supports.

SITING COMPOST PILES

The first thing to decide when making compost is where to site the piles. Access is needed so that compostable raw material can be delivered and taken away once composted. Allow space for several piles: material to be composted, material which is composting, material which has composted. Compost piles must usually be sited in such a way that all run-off (slurry) can be collected before it runs into the water table or local watercourse.

One practical reason compost is made above ground rather in subterranean trenches is ease of working. The key ideological reason, however, was explained by Rudolf Steiner.[6] He said that everything above ground level has 'a particular tendency to life, a tendency to become permeated with etheric vitality'. This applies both to crop plants like vines[7] as well as to compost piles. Thus the compost material is most easily permeated with the humus-creating and other formative forces the biodynamic compost preparations 502–507 bring if the pile is sited above ground level. Steiner says the compost should become 'inwardly alive' like plants which grow above ground level. As George Corrin says the aim of composting 'is not only to produce a *quantity* of humus to enliven our soil but also to produce humus of the right *quality*' [my italics].[8] Corrin's practical tips when composting include clearing grass away first to avoid a sour silage-like layer forming underneath the piles, siting piles on well-drained land and away from hollows where stagnant water might collect, and orienting them north–south so both sides of the heap are evenly warmed to ensure a more even fermentation, while allowing for both partial shade in summer and warming sun in winter.

Composting raw materials

Compost is made essentially from two main types of organic materials, those with a high nitrogen content and those with a high carbon content. The most common nitrogenous materials are fresh animal manures, fish wastes and freshly cut green plant materials (weeds, grass). The most common carbonaceous materials include hay, straw, shredded vine prunings, sawdust and other wood waste, dried seaweed and dead leaves. Nitrogenous materials are less stable than carbonaceous ones – piles of fresh animal manures left on their own tend to putrefy. Biodynamic composting rules only allow manure from animals on the holding or other biodynamic farms, or manure brought-in from extensive farming systems, meaning from livestock which is free-range. Until recently few wineries worldwide had any livestock of their own but this is changing, with horses, donkeys and mules (for traction), cows and alpacas (for manure), goats and sheep (for weeding) and ducks (for snail control) becoming more prevalent. Winegrowers can encourage local dairy farmers or shepherds to graze larger animals like cows on any spare land not given to vines; smaller animals like sheep often graze between vine rows over winter in New Zealand and parts of Australia, northern California, Provence and Italy. The stock save winegrowers the job of mechanically mowing the sward, allowing the land to enjoy a tractor-free rest in winter, and the animals find the varied and mixed herbage often present in bio vineyards to their liking. Because the animals graze the same land only in winter rather than year long there is no risk of parasite build-up either in soil or stock. However, stock with diets not approved by organic or biodynamic certifiers must first graze up to forty-eight hours in a quarantine paddock for residues from non-organic feed or antibiotics to be passed via the manure. Wineries in America (or those hoping to sell wine there) must remove stock (including chickens, geese, etc.) from vineyards ninety days before the grapes are picked under a recent USDA ruling, to reduce the risk of *E coli*.[9]

Winegrowers seeking Demeter biodynamic certification are actively encouraged to get some livestock onto the vineyard. Even if this may only be a few chickens to scratch around the vineyard to provide eggshells for Maria Thun's barrel compost 502–507 spray (or its derivatives), this can be a first step in allowing winegrowers to learn that, unlike vines, farm animals need housing and fencing plus daily feed and water. And, as New Zealand biodynamic consultant Bart Arnst makes clear, 'buying in manure for compost is one thing, but there is nothing like knowing the ingredients when baking your own cake'.

Cow manure is the best material as a base for compost, but other manures can be added in, such as chicken, horse, sheep or goat manures whose 'hot' character can be beneficial on colder (clay) soils, while 'cold' manure from ducks, pigs, and cows suits dry, sandy soil or places where it is too sunny.[10]

Which animal manure?

Joly says the choice of manure is critical because the animal's temperament is reflected in the quality of its manure, citing manure from the nervous, flighty horse as being especially hot.[11] Poultry and pig manures tend to be liquid and make unbalanced compost as far as vineyards are concerned, being excessively saline and rich in phosphorus and nitrogen, two elements vineyard soils rarely lack. Sheep and goat manures tend to be dry and as a result can heat poorly unless attentively composted, leaving a rather dry compost. Cow manure is the best all-round manure because generally it contains all the mineral and trace elements needed for wine production.

It has the perfect moisture level too, thanks to the huge quantities of saliva expressed during rumination and digestion by the cow, whose digestive tract is exceptionally long in proportion to its body compared to other animals or humans. The digestive fluids the cow manure contains are very beneficial for compost, a key reason why fresh manure is always best.

Poppen points out that 'the annual dropping of manure from a cow, properly handled, can make about four acres [1.6 hectares] of land fertile, while the same cow can live off approximately two acres [0.8 hectares]. It is this ability of the cow to give more than it receives that makes her so valuable.'[12] Spreading fresh, uncomposted manure on soil, however, is a mistake because while it is breaking down, the manure robs the soil of oxygen and releases ammonia nitrate. This results in problems similar to those caused by commercial fertilizers, such as burnt crop roots and excessively rapid plant growth from soluble nitrogen.

Carbon:nitrogen (C/N) ratio

Balanced compost is often principally defined by the ratio of carbonaceous and nitrogenous materials it contains, the so-called carbon:nitrogen or C/N ratio. Compost with a C/N ratio of 25:1 will contain twenty-five times as much carbon as nitrogen. Koepf provides examples of C/N ratios of compostable materials: sawdust is 150:1, straw is between 150 and 50:1, and manure with bedding material is between 25 and 20:1.[13] Koepf then states that a C/N ratio of between 25 and 30:1 is the ideal mixture from which to begin composting, and that finished compost should have a C/N ratio of between 14 and 20:1. He notes that stable humus in fertile soils has a C/N ratio of between 9 and 14:1.

As described above, the best nitrogenous material for composting is cow manure. Other common compostable nitrogenous materials include grass clippings, cut weeds or cover crops, but none of these are usually available to vineyards because after cutting they are invariably left as mulch to decompose back into the vineyard soil rather than be collected.

An important source of carbon for a vineyard compost pile is shredded prunings (if these can be collected from the vineyard, and the machinery for this task has improved dramatically of late) or other woody matter. Research suggests that prunings and woody matter, and also seaweed, help compost piles to develop organisms called actinomycetes. These filamentous bacteria resemble (and are sometimes wrongly described as) large diameter or long hyphae fungi. Actinomycetes give the characteristic earthy smell to soil and help maintain a 'forest floor'-type ecosystem there, one that some winegrowers argue is especially beneficial to vines which evolved along the edge of forests where they used trees as supports enabling them to climb in search of light (see Chapter 6, compost teas). The other advantage of carbonaceous matter is that it is more stable than nitrogenous matter, and if the compost pile smells of ammonia it means that nitrogen is being lost because manure in the pile has become too hot due to a lack of carbon to bind it. The pile is overheating. If there is a lack of nitrogen, however, the pile is too cool and will fail to ferment. A compost pile should smell only when the fresh manure it is being built from is being moved around. Once a built pile begins working (fermenting) it should do so in an almost odourless way.

Some growers add minerals to compost like slaked lime (calcium hydroxide) to lift soil pH (the lime should not come into direct contact with animal manure as this leads to rapid breakdown and a loss of nitrogen) and powdered rock (or pebble) phosphate or greensand to provide slow mineralization in phosphorus-deficient soils.

Just as basalt is added to barrel compost/cow pat pit 502–507 (see p. 107) it can also be added to compost piles, with Pierre Masson suggesting a dose of 2 to 5 kilos per cubic metre of compost. Alternatively, basalt can be sprinkled on animal bedding in the barn before this is composted.[14] However, making highly mineralized composts by 'treating the manure with all kinds of inorganic compounds and chemical elements … has no lasting positive effect,' said Steiner,[15] because like inorganic, chemical fertilizers they only affect the watery component of the soil.

BUILDING COMPOST HEAPS

Compost piles are long heaps called windrows. Standard practice is to make piles about 2 metres high and 1.5 metres wide at the base. The pile can be as long as available material and space allow. Heaps of different dimensions can create problems: a heap that is too big can exclude air and the breakdown can be anaerobic; a heap that is too small will dry out.[16] A permanent site will encourage a build-up of desirable organisms in the soil beneath, which

Compost piles in construction. A layer of grape stems from destemmed red grapes forms the base of the pile on the right and left (seen at the far end). The left-hand pile is also covered by a layer of grape pips and skins (bottom left) from red winemaking and a layer of cow manure and straw (middle). The finished piles each contained nine alternate layers in total.

can then quickly colonize any new heaps (this may entail installing an underground drainage collector for slurry run-off, and covering this with soil upon which the compost heaps will sit).

It is easier to achieve a homogeneous blend of carbonaceous and nitrogenous material in compost if alternate layers are made of animal manures and fresh green materials interspersed with straw, wood chips, shredded wood and so on, says Proctor.[17] He says to 'aim for 25 per cent animal manure content. If the manure is not available, make what you have into a slurry and water each layer with it. The layers of plant material should be 15 to 25 centimetres thick and those of animal manure no more than 7 centimetres.' Steiner suggested that around 10 per cent of any compost pile might also consist of added soil, topsoil rather than inert subsoil, or old compost.[18]

Andrew Lorand shapes his compost windrows into snake-like curves, avoiding the straight lines which he says are anathema to nature and which make it harder for the biodynamic compost preparations 502–

507 to complete their role of radiating etheric formative forces into the compost. Compost piles built in autumn are often covered with straw to keep the rain off and the cold out. Piles which start off too dry or become so will need watering. Piles which start off too wet will need turning. A well-made pile should need neither.

THE COMPOSTING PROCESS

The breakdown phase

Soon after the pile is built it begins to warm up in what is called the mesophilic or moderate-temperature phase. This first phase of breakdown lasts for a few days and is initiated by micro-organisms, mainly bacteria of the type commonly found in topsoil, which are mesophilic and capable of thriving at moderate temperatures of between 15 and 40ºC. They rapidly break down soluble, readily degradable compounds but also produce heat. The rise in temperature encourages thermophilic or heat-loving micro-organisms (mainly bacteria of the genus *Bacillus*) to take over the pile, precipitating the high-temperature composting phase. This can last from a few days to several months. Ideally the temperature at the heart of the pile will rise to 55°C, the point at which most micro-organisms pathogenic to humans or plants are killed, while staying below about 65°C so that the diversity of *bacilli* species which help decompose compostable material into stable humus remains.

The high-temperature phase is also known as the breakdown phase because this is when organic residues are decomposed into smaller particles by breakdown organisms, ammonifiers, nitrate formers and cellulose, sugar, and starch digesters. Ehrenfried Pfeiffer's work in identifying these breakdown organisms led him to develop his compost starter (see Chapter 5). Proteins are broken down into amino acids and amines, and finally to ammonia, nitrates, nitrites and free nitrogen. Urea, uric acids and other non-protein, nitrogen-containing compounds are reduced to ammonia, nitrites, nitrates and free nitrogen. Carbon compounds are oxidized to carbon dioxide (aerobic) or reduced to methane (anaerobic). As the supply of these high-energy compounds becomes exhausted, which may take from a few days to several months, the compost temperature gradually decreases and mesophilic micro-organisms once again take over and the final or build-up phase of the composting process begins.

In this phase simple compounds are resynthesized into complex humic substances. The organisms responsible for transformation to humus are aerobic nitrogen-fixing bacteria of the azotobacter and nitrosomonas group. Actinomycetes and streptomycetes also play an important role. Because soil is their usual habitat, adding around 10 per cent by volume of soil to piles, as Steiner suggested, favours their development and survival. Both actinomycetes and streptomycetes produce enzymes that allow them to break down tough woody debris. At this stage the compost is cooling but decomposition is not complete. Macro-organisms appear, of which the most important are two types of worm, earthworms (*Lumbricus terrestris*) and red wigglers (*Eisenia foetida*). Worms suck up bacteria, fungi, protozoa and organic matter while leaving castings which cleanse, feed and maintain soil. Worm activity is hindered in badly constructed piles where thick layers of loose, dry manure or straw create areas through which worms cannot penetrate. Such layers need moistening, consolidating and limiting in thickness.[19]

The build-up phase

The compost is finished when the development of humus can be seen because the compost pile has taken on a darker, more homogenous colour,[20] and the smell has changed from being raw and pungent to something more mature and earthy. Dig your hands into a mature compost pile to find out. Corrin says that the sign of well-made compost is if, when you press your foot against the side of the heap, 'you should have the sensation of putting your foot on a deep pile of carpet. Or pressing with your hand on the side of the heap should remind you of pressing your fist into a cow's side – yielding, neither dense nor loose.'[21] The moisture level of finished compost should be like a moist but not wet sponge, one whose moisture is felt but which does not drip even if you squeeze really hard.

Most winegrowers compost using the static pile method, in which compost windrows are built and left undisturbed for four to twelve months until needed. Industrial composting operations speed up the process, ventilating piles by piping in hot air or water and microbial inoculants. High-intensity windrows are subject to frequent turning. Whenever I use compost made using the latter two systems I recompost

*A purpose-built box holds the compost preparations before
each is inserted into the pile via its own separate hole*

it, incorporating it as a bulk filler in new static piles which also contain fresh material (manure, prunings, green waste, etc.) and which are seeded with the biodynamic compost preparations 502–507.

WHEN TO ADD THE BIODYNAMIC COMPOST PREPARATIONS 502–507

Usually the six biodynamic compost preparations 502–507 are added to compost piles immediately they are built. This ensures the beneficial forces the preparations carry can radiate throughout the pile's entire composting process, enabling life forces to be better retained, with less loss of nitrogen, and providing compost capable of promoting optimal crop growth.[22]

Growers who liken certain compost preparations to bodily organs – oak bark 505 to an organ of thinking, i.e. the brain; stinging nettle 504 to the heart; chamomile 503 to an organ of digestion – are especially keen to put the preparations in compost piles immediately they are built.[23] In this way the compost itself becomes a living organism. Those biodynamic growers who fear the effect the preparations have may be compromised by too much early heat in the pile during the high-temperature phase may decide to add the compost preparations only after the temperature of the pile first peaks, when it has cooled to 35.0 to 37.5°C.[24] However, this delay may weaken the 'biodynamic effect' of the compost.[25] Maria Thun says adding the biodynamic compost preparations to a newly built pile produces a cooler initial fermentation.[26] My own view is that if the biodynamic compost preparations are to hand when you make the piles then put them in – but if not, wait until the pile's temperature peaks and then add them. If you are worried or unsure about the effect excess heat might have on the preparations then put two sets in, one as you build the pile and another after it has cooled down or when (and if) you turn the pile.

HOW TO ADD THE BIODYNAMIC COMPOST PREPARATIONS 502–507

The five solid biodynamic compost preparations – yarrow 502, chamomile 503, stinging nettle 504, oak bark 505 and dandelion 506 – are usually dropped into the centre of the pile via holes made at regular intervals along either the sides or top. The holes can be made using a wooden post or rake handle. Solid compost preparations can become stuck halfway down what are deep, narrow holes and thus fail to reach the centre of the pile. To stop this from happening first roll each one into a separate ball or sausage of earth or old compost before dropping this into the hole and backfilling it. This method also helps prevent preparations made in a drier, flakier (less humic) style from blowing away in windy weather as they are being dropped in. The liquid valerian 507 preparation can be sprayed over the entire outside layer of the compost windrow, or half can be sprayed over the top and the remainder can be poured directly into a separate hole made in the pile.

Lovel argues that both the location of the holes in the pile and which hole each preparation is placed into are significant.[27] He says two holes should be made an equal distance apart and right in the deep centre of the pile for the stinging nettle 504 and oak bark 505 preparations. To Lovel these preparations represent the sun and moon. They are made from leaves, stems or bark, meaning parts of the plant which draw energy into the plant and keep it there. Their role in compost is to draw etheric formative forces into the pile and keep them there. The other four holes are made around the edge of the pile but are less deep than the previous two. These holes are for the yarrow 502, chamomile 503, dandelion 506 and valerian 507 preparations which respectively represent Venus, Mercury, Jupiter and Mars. These four preparations are made from flowers, the plant organ which releases energy. What is released is then drawn into the centre of the pile by the stinging nettle 504 and the oak bark 505. Lovel adds that yarrow 502 and valerian 507 work best as a pair sited opposite each other, as do chamomile 503 and dandelion 506.

In line with the biodynamic idea of all preparations carrying etheric formative forces, the quantities of the biodynamic compost preparations 502–507 are less important than the qualities they bring. In any event only tiny amounts of compost preparations are added to compost piles: 1 to 2 square centimetres of each (roughly a level teaspoon) of the six compost preparations should be added for every 10 square metres of compost material.[28]

Calculating the weight of finished compost is tricky but easier than calculating the weight of pre-composted material, because finished compost should have a more homogeneous texture and moisture level. My rough calculation when making vineyard compost is that 1.5 cubic metres of fresh pomace, winter barn manure from horses and cows which is no more than six months old (so still warm and moist) and pre-composted, dryish vegetation from a large organic garden will, when combined, compost down to 1.1 cubic metres or roughly one metric tonne of biodynamic compost with a margin of error of 10 to 20 per cent. Ten cubic metres of compost is a pile 1.6 metres high, 1.2 metres wide and 5.2 metres long.

SPREADING BIODYNAMIC COMPOST 502–507

An old saying suggests that good compost can be applied at any time on any crop and in any amount.[29] When spreading compost, winegrowers should aim to leave soils with a stable total organic matter content of 1.7 to 2.5 per cent, the exact level depending on soil type.[30] Compost for vineyards is usually spread in autumn, after harvest and before pruning begins, when the earth is breathing in and the soil is still warm enough for bacteria and other micro-organisms to incorporate it. The compost can be left to decompose on the surface of the soil but is more usually turned in using disks and to a shallow depth. Too deep and it may turn anaerobic.

As spreading compost invariably involves moving heavy machinery in and around vineyards (tractors, compost spreading trailers), spread compost occasionally but in generous doses: from 5 to 15 tonnes per hectare every two to five years for healthy vineyards on good soils, up to 15 to 40 tonnes per hectare every one to four years for weak vines on poor sandy soils lacking humus because they leech nutrients quickly. The highest doses are reserved for conventional vineyards converting to organics/biodynamics and which suffer from eroded, compacted soils and contain vines suffering esca and other vine cancers. The life forces which compost brings to the soil, allied to better vineyard hygiene, can greatly slow the rate at which such illnesses spread without, of course, actually curing the affected vines.

In hot climates compost is often most efficiently spread as an under-vine mulch. Required volumes are between 100 to 300 cubic metres per hectare depending on vine density. After spreading, the compost is covered with straw mulch to extend its *in-situ* lifespan, which is normally around two to three years. Costs of mulching are similar to those had the vineyard been weedkilled for the same period of time. Prue Henschke of Henschke (www.henschke.com.au) in South Australia's Eden Valley says the advantages of mulching over weedkillers are that 'the soils hold more nutrients, stay moist in drought conditions and let vines take up water more slowly. Overall our vine growth is now more regular, analysis shows the vines are less stressed and we get no berry shrivel.'

Graeme Sait (see p.xvi) says, 'I have yet to meet a viticulturist who has used compost who does not continue with this practice. Compost adds humus, a new microbe workforce and complex minerals beneath the vines. It also triggers more humus formation and can promote a resilient soil, particularly if it contains predatory fungi.'

4

DYNAMIZING (STIRRING)

The underlying principle of the nine biodynamic preparations 500–508 is that they are physical substances which carry intangible etheric formative forces. The forces contained in the six compost preparations 502–507 reach large areas of farmland when biodynamic compost is spread. As well as bringing forces, the compost also brings to the land valuable physical substance in the form of organic matter, nutrients and living organisms. However, for the forces contained in the three biodynamic field spray preparations – horn manure 500, horn silica 501 and common horsetail 508 – to reach the farm it is first necessary to dilute them in water and then stir or 'dynamize' them. The same dynamizing process is also used for the only liquid compost preparation, valerian 507, before it is applied to the compost.

On one level, the stirring or dynamizing process allows both the forces contained within the preparations as well as those from the wider celestial sphere to reach the land. The theory is that water keeps the memory of the dissolved biodynamic preparation,[1] and that this 'information' can be transferred.[2] On a practical level, stirring helps the substances to be thoroughly mixed in the water. The oxygenating effect the stirring has brings a substantial increase of oxygen in the water, up to 75 per cent after one hour of manual stirring, according to Pfeiffer.[3] This helps microbes present in, for example, the horn manure 500 or Maria Thun's barrel compost 502–507 sprays to multiply rapidly.

THE SIGNIFICANCE OF THE VORTEX

Steiner described the stirring or dynamizing process he intended when discussing horn manure 500: 'You must make sure ... that the entire contents of the horn have been thoroughly exposed to the water. To do this, you have to start stirring it quickly around the edge of the bucket, on the periphery, until a crater forms that reaches nearly to the bottom, so that everything is rotating rapidly. Then you reverse the direction quickly, so that everything seethes and starts to swirl in the opposite direction. If you continue doing this for an hour, you will get it thoroughly mixed.'[4]

When stirring a biodynamic preparation in water in a vertical container a whirlpool effect is created by the wall of water which forms at the centre. Steiner described this as a 'crater', but this is now more commonly referred to as the vortex. The vortex should be as deep as possible, going to the bottom of the container.[5]

Jennifer Greene calls the vortex the water's sense organ.[6] This recalls the idea of how six of the nine biodynamic preparations are sensitized to formative forces by being enclosed in sense organs (animal sheaths). The vortex is the water's way of rhythmically ensheathing the forces contained in the preparations. When the direction of the stirring changes from one way to another the vortex is lost as the water seethes and undergoes chaos. This moment of chaos is when the preparation being stirred is said to receive the imprint of the cosmos, whose forces stream into the earth leaving their imprint on all living things.

Greene worked with Theodor Schwenk, a German hydroengineer and pioneer in water-flow research whose book on the subject, *Sensitive Chaos*,[7] was described by Commandant Jacques Cousteau as the first phenomenological treatise on water. Schwenk identified the vortex as the living pulse which allowed water to take in air in such a way that the water could regenerate itself – just as it did in natural springs – and stay cleaner and fresher, not simply to purify it but to revitalize it as well so that it could then support living processes. Hence Greene says that while we happily think of water in terms of pollution or its use as a mechanical or engineering (hydraulics) aid, water as an element for the purveyor of life has not generally been the modern focus.[8] Nicolas Joly reminds us that the latent forces underlying life are often manifested in physical

A vortex, water's 'living pulse', in a stirring machine made in New Zealand from an old milk storage vessel

matter in the form of spirals: calving rings on vortical shaped female cow horns are the prime biodynamic example (see Chapter 2, horn manure 500).[9] Another might be our universe, a spiral in which our planets constantly swirl. Spirals of water are what form or create the vortex.

Greene describes how Schwenk noticed that movement at the centre of a funnel or vertical vortex is faster than the movement on the periphery, and that this exemplifies Kepler's Laws of Movement observed with the planets: those closest to the sun move more quickly than those furthest away.[10] However, if you take a small rectangular piece of paper, put a dot at one end and place this piece of paper midway between the periphery and the centre of the water and air surface, the paper will move in a such a way that it maintains the same orientation. As Schwenk describes it, it retains its orientation to a fixed star. The conclusion drawn is '*that* water moves pertains to earthly forces' but '*how* water moves' pertains to cosmic laws.

West argues that spraying herb teas and liquid manures like those described above without first dynamizing them encourages plants to feed directly off the substances they contain through their water roots, exactly as if inorganic water-soluble or 'chemical' fertilizer was being applied.[11]

The act of dynamizing and the vortex which the stirring process creates mean nutrients suspended in the water carry an electrical charge. This renders them colloidal, and for plants to feed naturally nutrients must be colloidally bonded to an organic molecule. Plants fed with an organic colloidal system rather than an inorganic water-soluble one have more feeder root hairs and thus stronger, healthier root systems because this is exactly how nutrients pass from soil solids to plants anyway. Through the chelating action of the aerobic bacteria, nutrients are in the perfect form for a plant to utilize and have become part of a living organism, namely the soil. Plants stay healthier when they have more food to choose from because nutrients held as liquid colloids do not leach from the soil when it rains in the way that water-soluble fertilizers do. By implication, colloidally held nutrients will provide greater health and vitality to humans hoping to assimilate nutrients from crops they consume.

The container for dynamizing

A common container used by winegrowers for stirring is an old barrel with one of the ends removed. The barrel can be cleaned naturally by weathering it outside rather than scrubbing it with detergent or charcoal. Those who favour containers made of inert materials should try to avoid galvanized vessels (especially those with chipped surfaces which encourage rust), and if there is no alternative to plastic one of the hard, dense types should be chosen,[12] presumably to reduce the risk of off-gassing. Pottery is another option. For stainless-steel containers, non-magnetic forms should be used. The stirring vessel should be sited in a place which makes use of gravity both for filling it with water and draining dynamized liquids ready for spraying. In terms of its dimensions, Pierre Masson suggests the dynamizer should be taller than it is wide, giving 1.4:1 as an ideal ratio, and with a rounded out bottom for a better stirring effect.[13]

Dynamizing by hand

Ideally every biodynamic farm would be of a size which permitted all tasks to be carried out manually, even stirring, although Steiner acknowledged that for larger farms mechanization would be necessary.[14] Yet Steiner was clear which method was preferable, saying, 'there's no question that stirring by hand has a quite different significance than mechanical stirring, although of course someone with a mechanistic world-view would never admit it.

Just consider what a huge difference there really is: when you stir by hand, all the fine movements of your hand go into the stirring, and quite possibly all kinds of other things do too, including the feelings you have as you stir [namely the importance your will has in manifesting the spirit in physical life. See scientific substance versus spiritual force, above]. People nowadays don't think that makes any difference, but in the field of medicine, for instance, the difference is quite noticeable. Believe me, it is really not a matter of indifference whether a certain medication is prepared by hand or by machine. Something is imparted to the things that are produced by hand.'[15]

For this reason it is recommended that only one person should stir any given preparation, rather than have several people take turns stirring the same preparation for short periods; that once started a dynamization should never be interrupted; and that a dynamized solution should never be mixed, either with another dynamized solution (see Chapter 2, horn silica 501) or a non-dynamized one.

Germany's Institute for Biodynamic Research (IBDF) has changed its advice on the application of biodynamic spray preparations in recent years through its research into formative forces, now recommending hand rather than machine stirring.[16]

Storch makes an observation few involved in biodynamics would disagree with: 'Anyone who has done a serious amount of hand stirring will have noticed a point in the stirring process where there is a transformation in the liquid. The stirring gets easier and there is a noticeable difference in the ease with which the vortex forms and there is something in the stirred liquid that changes that I cannot put my finger on.'[17] The likely answer is the oxygenating effect the stirring has – although Rhône-based winemaker Michel Chapoutier suggests the water's change in texture arises because dynamizing splits clusters of water molecules, energizing them via the action the vortex has on quarks, quarks being hypothetical elementary particles at a sub-atomic level.[18]

Peter Proctor says that when the water becomes more slippery and viscous and easier to stir, then 'the water has become enlivened by a similar process to that of the growing plant, the rhythm of the expansion to leaf and contraction to seed. In this process you have increased the oxygen content of the water. At the same time you have introduced the cosmic forces that enable the water to become a dynamic carrier of

Bertie Eden of Château Maris in the Minervois,
France and his wooden dynamizer

the life energy of the [biodynamic preparation] as it is spread over the land.'[19]

Hand stirring can involve either placing your hand directly in the water-filled recipient, or using your hands to move an implement, such as a pole suspended in the water from above (attached to an overhead beam, for example), which makes stirring larger volumes of water less tiring.

Dynamizing by machine

The number of companies supplying stirring machines aimed at the biodynamic market is small but increasing. Suppliers I am aware of include Matthieu Bouchet's Terres en Devenir in Montreuil-Bellay (Maine et Loire), France, for dynamizers made from wood (see photo of Robert Eden, above) and latterly from clay-lime (and thus metal-free) mortar. Gian Zefferino Montanari of Bio-Meccanica (www.biomeccanica.com) in Scandiano (RE), Italy, produces cleverly designed and mechanically robust copper stirring machines and sprayers framed in stainless steel. The electric motor is insulated and moveable and so can be sited away from the machine when a dynamization is taking place. Unframed alternatives, also from copper, are made by the Swiss Ulrich Schreier of Eco-Dyn (www.eco-dyn.com)

in Becon Les Granits (Maine et Loire), France, now by far France's leading supplier of copper dynamizers to wine estates. Steve Storch (www. naturalscienceorganics.com) in Water Mill (NY), US, designs and produces barrel-shaped copper stirring machines which are hydraulically rather than electrically powered. As he says, it makes no sense going to the trouble of storing and keeping the biodynamic preparations away from electricity if at the most intense moment they experience when being stirred in water they are blasted with electromagnetic fields from electrical machinery.[20]

There are two possibilities as to the moment when the reversal in direction of the movement of stirring which creates chaos should take place. One is that it should be decided by the height the water reaches as the central vortex pushes the water up around the side of the stirring container. This takes account of the suppleness of the water, and how it changes during stirring (see Storch's comments above). The other possibility is deliberately pre-programming the change in direction every twenty to thirty seconds or so, with a short pause. Placing the drive mechanism which turns the stirring paddles beneath the chamber holding the water, rather than above it, may allow even better penetration of celestial forces. Biodynamic growers are encouraged to

Italy's Gian Zefferino Montanari builds dynamizers (behind his right shoulder) and dedicated sprayers for both the biodynamic horn sprays (left shoulder) from copper

remain near mechanical stirring machines when they are being used so that 'the intention of the farmer [remains] fully involved with the stirring',[21] meaning that even if the machine is a money-saving device it should not be seen as merely a time-saving device as well. This echoes Steiner, who said that 'you may say that enthusiasm cannot be weighed or measured, but an enthusiastic doctor is an inspired doctor, and the doctor's enthusiasm supports the effect of the medicine'.[22] The words 'farmer' instead of 'doctor' and 'farm spray' instead of 'medicine' can be used for his point to have an agricultural resonance.

FLOWFORMS

The traditional way of dynamizing using a bucket, tub or tank results in a single central vertical vortex or whirl forming in the water. At most around 200 litres can be stirred by hand in this way, whereas volumes of around 600 litres can be stirred in the largest mechanically powered vertical dynamizers. Much larger quantities of water – up to several thousand litres – can be dynamized by using what are called flowforms. The first flowforms were designed in the early 1970s when Theodor Schwenk asked John Wilkes (1930–2011), an English potter turned sculptor and graduate of the Royal College of Art, to make a model over which water could flow.

Wilkes wanted to work on the resistance in streaming water and its resulting rhythms. During his experiments, he found that, by creating a certain resistance to the water flow in a vessel with defined proportions, a pulsing figure-of-eight pattern would arise. Thus the flowform principle was discovered. The work into what became known as flowforms and how stirring affects the biodynamic preparations took (and still takes) place at Herrischreid in Germany at Schwenk's Institut für Strömungswissenschaften or Institute for Flow Sciences (www.stroemungsinstitut.de). Wilkes installed his first flowform in 1971 in Järna, Sweden. This prototype was later commercialized as the Järna model.[23]

A flowform consists of a series of between three and up to a dozen or more small and symmetrical sculpted vessels or basins through which water is channelled. Water flow rates vary between 1,000 litres per hour, for a three-basin flowform, up to 2,800 litres per hour for a seven-basin flowform. Each basin has an inlet and an outlet. Water is channelled to cascade and stream through these sculpted forms in a way which

Matthew Frey and flowform at Frey Vineyards,
Mendocino County, California

replicates certain archetypal forms found in water's movement in its natural state, such as if it were eddying and flowing over pebbles in a stream. The proportions of the basins create and maintain a rhythmical pulsing figure of eight or lemniscatory movement in the water which replicates how blood moves in living organisms. Wilkes said this figure of eight was fundamental to organic processes.[24] The water undergoes chaos when it falls in a collecting tank and can then be returned to the top of the cascade for the cycle to begin again.

Flowforms create layers in the water because streams of water are moving in the same direction in the basins, but at differing speeds. This allows multiple vortices to develop, in contrast to the single vertical vortex produced in a vertical container. When two layers of water flow past each other creating resistance, planes of vortices form between these layers. This dramatically increases oxygen levels in the water and does so in a rhythmical way which is said to enhance the water's vitality. As well as being used in biodynamics, flowforms are also used to treat industrial and agricultural effluent because, as initial research at the Rudolf Steiner Seminariat wastewater treatment ponds in Sweden showed, they can oxygenate large volumes of water while also stimulating biological activity. Winery waste water can be passed through a flowform to clean and revitalize it. The visual and aural effect flowforms have mean they also have a role in therapeutics.

Research into flowforms is being undertaken at the Michael Fields Research Institute (www.michaelfields.org) in Wisconsin, US.

THE WATER

The effectiveness of water-based sprays may be improved by bearing in mind some of the following observations as regards water quality.

Tap or mains water is deemed unsuitable for biodynamic spray preparations, teas, liquid manures and decoctions. It is too hard (alkaline) and contains added fluoride and chlorine, as well as nitrate (fertilizer) and pesticide residues from agricultural run-off. Although these cannot be removed, chlorine levels can be reduced by aerating the water either by passing it through a hose with a sprinkler head or splashing it into a container which is left uncovered.

Pierre Masson favours water with a slightly acidic pH of between 6 to 6.5 and with a low mineral content, like rainwater.[25] To collect it, let it rain for twenty minutes before placing a recipient container under a roof gutter because the earliest fraction will be the most polluted and contain mineral impurities which may inhibit the fermentation process for liquid manures. It is also likely to contain accumulated dust and other dirt from the roof itself.

The pH of alkaline spring or river water can be lowered (acidified) with cider or vine vinegar, while that from acidic, granite-rich soils can be raised by adding calcium-rich marine maerl (calcified sea algae).[26]

SPRAYING DYNAMIZED LIQUIDS BY MACHINE

When Steiner was asked, in the discussion after the fourth lecture of his 1924 *Agriculture* course, whether the forces carried by a dynamized liquid would be lost by using a machine that breaks the liquid up into a very fine spray he replied, 'Not at all. They are very firmly bound. In general, you don't have to be nearly as afraid that spiritual things will run away from you as you do with material things.'[27]

5

OTHER BIODYNAMIC SPRAYS AND TECHNIQUES

In March 1925, less than a year after giving his *Agriculture* course, Rudolf Steiner died. Before his death Steiner maintained that in an age of increasing empiricism it was perfectly natural that the spiritual and scientific indications the course contained should be verified by testing. A *Versuchsring* or Experimental Circle was established, supported by the Faculty of Science at the Goetheanum (see Appendix II). Initially research focused on proving Steiner's biodynamic preparations did have measurable effects. Since then various biodynamic practitioners worldwide have refined Steiner's biodynamic indications or developed wholly new ones for sprays and spraying, which this chapter covers, as well as for celestial forces which are covered in Chapter 7.

In 1928 sixty-six farms were using what had become known as the Biological-Dynamic Method, or biodynamics. Food produced using biodynamic methods now had its own Demeter trademark (see Chapter 8), while the Experimental Circle had 148 members.[1]

After Steiner's death Ehrenfried Pfeiffer (1899–1961), the German biochemist Steiner had chosen to make trial batches of biodynamic preparations for the *Agriculture* course, became the leading figure in the biodynamic movement. Pfeiffer helped develop biodynamic farms in Germany, Switzerland and the Netherlands before moving to Pennsylvania in 1938, not long before the Nazi regime outlawed biodynamics in Germany. Other key German biodynamic figures, like Eugen and Lili Kolisko (see Chapter 7), had already left Germany, continuing their biodynamic work in the UK.

In the US, Pfeiffer established a model biodynamic farm and founded the Biodynamic Farming and Gardening Association of North America (www.biodynamics.com). Pfeiffer's work, which is continued by the Pfeiffer Center (www.pfeiffercenter.org) in Chestnut Ridge, NY, included developing a diagnostic technique called sensitive crystallization plus two proprietary sprays, the Pfeiffer Compost Starter and the Pfeiffer Field Spray Concentrate, which are dealt with immediately below.

SENSITIVE CRYSTALLIZATION

This analytical technique peculiar to biodynamics came about when Rudolf Steiner asked Dr Ehrenfried Pfeiffer to find a reagent that would allow the 'formative forces' of a living organism or system to be revealed pictorially, so that the effectiveness or otherwise of biodynamic trials could be gauged. The technique works by allowing copper chloride crystals to form patterns, resembling ice crystals on a window pane, when a water-soluble extract of a

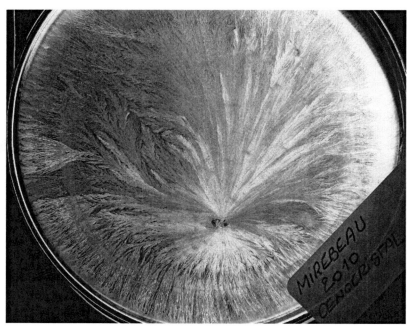

Sensitive crystalization slide made of a 2010 Graves Rouge from Château Mirebeau in Martillac, Bordeaux by Margarethe Chapelle of the Centre de Recherches et d'Études Qualitatives (Oenocristal), Puy L'Evéque, France

substance like wine, plant sap or milk is added to it. Both the structure of the crystals and their texture make visible the 'inner quality' or etheric formative forces (or lack thereof) in the material being tested. Some contemporary winegrowers use sensitive crystallization as one way of evaluating the structural and organoleptic characteristics of wine quality.[2]

PFEIFFER COMPOST STARTER

Pfeiffer designed his compost starter for industrial and urban waste, trialling it commercially between 1950 and 1952 whilst directing a municipal composting programme in Oakland, California. Household waste was composted then pelletized for use as agricultural fertilizer. Based on the six biodynamic compost preparations 502–507 and biodynamic horn manure 500, Pfeiffer's compost starter also contained bacteria, fungi, actinomycetes, yeasts and other micro-organisms capable of generating large quantities of stable compost from green waste via a quick, hot compost fermentation.[3] Having worked with Steiner, Pfeiffer knew that inoculating compost with a microbial starter was no magic bullet as this encourages a materialistic (physical substances) rather than spiritual (intangible forces) view of the composting process. Pfeiffer's book, *Bio-Dynamics – A Short Practical Introduction* which was published in 1956, presented biodynamics in a down to earth manner. No explicit reference was made to the strange ways the biodynamic preparations were made, presumably so as not to scare off American farmers with the no-nonsense mindset of the 1940s and 1950s.[4] Pfeiffer's proprietary sprays got farmers using the biodynamic compost preparations 502–507 without having to endure the goriness of making them. The downside is that off-the-shelf products like Pfeiffer's allow farmers to avoid adopting the biodynamic mindset of working things out for themselves.[5]

PFEIFFER FIELD SPRAY CONCENTRATE

Pfeiffer's second spray, a field spray concentrate, was made in a similar way to the compost starter but was intended to be sprayed directly on farmland. Its role was to build humus and help stubble or cover crops (green manures) to break down before they were ploughed into the soil.[6] Two variations of Pfeiffer's field spray concentrate include Maria Thun's barrel compost spray 502–507 and Peter Proctor's cow pat pit spray (CPP) 502–507.

MARIA THUN'S BARREL COMPOST SPRAY 502–507

This is simply a speeded up form of solid biodynamic compost 502–507 applied in infinitely smaller volumes and in liquid form. In Germany it is called *Pfladenpräparat* ('cow pat preparation'), in France it is called *le compost de bouse*, and in Australasia a variant of it made by Peter Proctor using a brick rather than barrel-lined pit is called cow pat pit (see below). Barrel compost has numerous other names – barrel prep, barrel manure, biodynamic compound prep, dung compost spray, manure concentrate – but has one main function, offering an easier, quicker way of getting the biodynamic compost preparations 502–507 onto vineyards rather than waiting the six to twelve months compost piles need to mature. This is especially appealing to winegrowers with large or steep vineyards, or who are unable to find enough of the right compostable material.

The barrel compost spray 502–507 is seen as a helpful primer for land converting from conventional farming to biodynamics, preceding even the very first application of horn manure 500, and initiating healing processes for soils by reversing the erosive, hardening tendency soluble fertilizers have of turning clay (aluminium silicate) back towards rock.[7]

Barrel compost should not be viewed as a long-term substitute for solid biodynamic compost which imparts more profound, longer-lasting effects on both soil structure (substances) and soil vitality (forces). Barrel compost 502–507 softens compact soil by bringing air into it, balancing soil nutrients, improving soil structure, decomposing organic matter (e.g. cover crops), stimulating humus formation and generally improving soil quality. It can also be sprayed on animal bedding before this is composted.[8]

Barrel compost was developed in the early 1970s by Maria Thun (see Chapter 7).[9] Its precursor was the 'collective preparation' or *Sammelpreparat* developed in 1927 by Max Karl Schwarz, one of the first German farmers to adopt Rudolf Steiner's biodynamic ideas. Schwarz made his collective or 'birch pit' preparation by lining a long pit dug into the ground with birch poles and locating it near the barn in which farm animals overwintered. Their manure was emptied into the pit and sets of the biodynamic compost preparations 502–507 were dropped in. The manure was left to compost for several months and then spread on the farm in solid form, and the pit was refilled with fresh manure.

Rudolf Steiner made no reference either to barrel compost or a collective/birch pit preparation in his 1924 *Agriculture* course.

Making Maria Thun's barrel compost 502–507

The main constituent of Maria Thun's barrel compost is fresh cow manure. This can be collected in the same way as for horn manure 500, although just before her death Thun suggested manure from pregnant cows be used.[10] The pats can be up to two days old.

To 50 litres of cow dung, add 500 grams of basalt as coarse grains, grit or finer powder and 100 grams of finely crushed, sun-dried (not oven-dried) eggshells. These are then mixed together for an hour. Using

Flicking freshly stirred (dynamized) valerian 507 over manure is the final task when making Maria Thun's barrel compost 502–507

shovels, this can be done either in a wheelbarrow or in a wooden barrel stood on one end with the other knocked out, or on a board placed on the ground. Stir from the outside in.

Thun says the mixture should have become 'one dynamic whole', resembling a big cow pat with a slightly dilute colour. Peter Proctor says farmers can become bored during the mixing or stirring and not mix as well or long as required, but a good stirring will 'make all the difference' to this preparation's quality. Others say using a cement mixer is easier still...

The mixing can be carried out under a descending moon and when the sidereal moon stands in a root/earth constellation: Virgin in the northern hemisphere and either Goat or Bull in the southern hemisphere.

Half of the manure, basalt and eggshell mixture is then placed in another barrel stood on one end, but with both ends knocked out. This would have previously been dug into a hole in the ground, not quite half as deep as the barrel, with the excavated earth piled around the part of the barrel poking up above ground level. The barrel is left open at both ends so the contents within may receive both earthly (lime/calcium) and celestial (silica) forces. The five solid biodynamic compost preparations 502–506 are dropped one by one and separately into the mixture, with stinging nettle 504 usually placed at the centre. Then the remaining half of the manure, basalt and eggshell mixture is placed on top, and it too has a set of solid compost preparations inserted. Finally, a liquid mixture made from five drops of the valerian 507 preparation stirred for ten minutes in a litre of water is poured over the top.[11] The barrel is then covered with its lid.

Twenty-seven days or one sidereal month later, when the descending, sidereal moon has returned to the same earth/root sign under which the mix was prepared and added to the barrel, the barrel's contents are aired by turning them briefly with a spade. Thun says that after another two weeks the barrel compost will be ready; Bouchet says to wait another sidereal month. Some growers leave their barrel compost in the barrel for several more months or until it is needed for use, adding an extra set of preparations whenever the barrel compost is turned. Some New Zealand growers leave their barrel compost (or, more usually in their case, Peter Proctor's cow pat pit (CPP) Spray 502–507) up to one year for it to experience all four seasons.[12]

Leaving the finished preparation *in situ* risks allowing worms to devour it, however. The finished preparation should resemble very rich, dark, fine soil with a clean and intensely earthy smell. It is stored in the same way as horn manure 500.

The role of eggshells and basalt

Thun decided to add calcium-rich eggshells to her barrel compost because she found that oats, celery and tomatoes grown on limestone soils had healthier root systems and contained fewer residues of the radioactive Strontium 90 left by America's 1958 atomic bomb tests compared to similar plants grown on sandy (siliceous) granitic soils. Biodynamic winegrowers are therefore encouraged to keep chickens from which to source fresh eggshells for this part of the preparation.[13] Thun says basalt's role is to support those living organisms and processes in the soil involved in or which work towards decomposition, and thus promote humus formation. Henderson points out that in chemical terms this leads to the formation of more clay minerals, which encourage humus formation (clay-humus complex).[14] The basalt acts in a nitrogen-fixing capacity when even more finely ground than grit, she says.

Another way of looking at the addition of eggshell (calcium) and basalt is that they represent two basic soil types: basalt is of ancient volcanic origin and comes from inert magma within the earth's mantle and amounts to embryonic, new soil,[15] like infant clay;[16] calcium is a geological baby present in limestone-rich soils formed by marine deposits from living creatures within the last several hundred million years. From a biodynamic perspective the eggshells provide the lime polarity (crop growth) while basalt provides the balancing silica polarity (crop taste). Christian von Wistinghausen used volcanic ash instead of eggshells for his version of Maria Thun's barrel compost, the *Mäusdorfer Rottelenker,* which he named after his home town of Mäusdorf.

A role for stinging nettle?

Adding finely chopped (2.5 centimetre) fresh stinging nettles to the manure mix appears to regulate plant health and growth, possibly by stimulating greater root growth and improved root health. The nettle should be harvested as it is about to flower.[17]

Peter Proctor's cow pat pit (CPP) spray 502–507

Peter Proctor called his version of Maria Thun's barrel compost the cow pat pit or cow pat prep ('CPP'). This is because a shallow pit or trench about 90 centimetres long by 60 centimetres wide and 30 centimetres deep is used, rather than a barrel. Proctor found making barrel compost problematic 'because it is slow, and hard to get the preparation out of the barrel when it is ready; and the preparation can smell because it has gone anaerobic in the barrel. I find it easier to make the preparation in a pit lined on all four sides by old bricks. These absorb moisture but keep the dung cool, while stopping it from drying out.'

Proctor lays the dung to a depth of 10 to 12 centimetres. 'Any deeper and the transformation process will take too long. It should take about two months,' he says. He lines the pit with the cow dung together with 200 grams of powdered eggshell and 200 grams of basalt dust. 'These are first mixed together for fifteen to thirty minutes in your hand,' he says 'rather as you would mix dough, with a sort of flipping motion. Put the mixture into the pit and pat it down, but pat lightly, as you do not want to overly compact the mixture. It should be level. Then add 1 to 3 grams of the solid biodynamic compost preparations [502–506]. Then

Peter Proctor shows off a brick pit for his cow pat pit/prep in his back garden in New Zealand

potentize [stir] the valerian compost preparation [507] in the usual way and sprinkle over the cow pat mixture. Then spread a damp hessian sack or gunny bag over the top. You should lay the brick pit in a shady place to keep it cool by allowing a good air flow in hot weather, to stop it from getting wet from rain, and to keep it sheltered in cold weather. The aim is to achieve a constant temperature and humidity. So in dry weather you can sprinkle the bricks with water every two or three days to keep them damp and to maintain humidity. After six weeks turn the preparation with a garden fork. If the dung has not broken down add another set of compost preparations. The dung should lose its smell. You will see decomposition begins at the edges of the pit where air flow is greatest.'

Spraying barrel compost/cow pat pit 502-507

Before being sprayed on the land, barrel compost/CPP is diluted in water and dynamized for twenty minutes, rather than for a full hour which is the case for the two biodynamic horn preparations 500 and 501. This is because the manure in the barrel compost/CPP has already been mixed for

A variation on Maria Thun's barrel compost 502–507, using Peter Proctor's idea of bricks instead of a wooden barrel at the Josephine Porter Institute for Applied Bio-Dynamics, United States

one hour in its solid form, so only twenty minutes of stirring is needed when the finished preparation is diluted in water prior to spraying. If the compost preparations 502–507, already present within barrel compost/CPP, were to be dynamized for another full hour certain beneficial processes risk being inverted, such as barrel compost's anti-cryptogamic effect.[18]

For one hectare, Thun mixes 240 grams of her barrel compost in 40 litres of water. The diluted preparation is applied as a fine spray on the soil within at most four hours of dynamizing. For one hectare, Proctor dilutes 2.5 kilos of his CPP preparation in 112 litres of water, over ten times as much as Thun, probably because Proctor's main consulting work was with Indian farmers working parched, easily eroded soils. The much more concentrated dilution means Proctor's CPP acts almost as a liquid soil manure, or manure concentrate.[19] Cow pat pit is the most popular form of barrel compost in Proctor's native New Zealand.

The effects of barrel compost/cow pat pit 502-507

Barrel compost/CPP's role is to activate soil organisms while improving soil structure. It is commonly sprayed on freshly ploughed soil or on soil about to be ploughed, such as when compost or green manures (cover crops) are being turned in,[20] usually either in autumn or spring. Spraying barrel compost in autumn stimulates soil microbes just as the vine roots they often colonize become active and fill with carbohydrates from the previous year's vine growth, the very food these microbes not only need but have been waiting for. A follow-up application of barrel compost/CPP in spring is seen to bring to a close the winter process of decomposition underground. By bud burst in spring, soil organisms should have finished decomposing compost, cover crops, fallen leaves or vine prunings left between the vine rows. If this process is incomplete then vines may struggle to find the soil nitrogen they need for bud burst and new shoot growth.

Like horn manure 500, barrel compost/CPP 502–507 is most effective when sprayed in the afternoon, in autumn and under a descending moon when the earth inwardly inhales. Mixing barrel compost/CPP 502–507 and horn manure 500 together to save spraying time risks creating the same opposition of forces potentially present in prepared horn manure 500 + 502–507. This is why Bouchet advises leaving a three-week gap between spraying horn manure 500 and barrel compost

502–507 (see also winter tree paste, below).[21] Proctor advises waiting at least two days after spraying CPP 502–507 before spraying either horn manure 500 or horn silica 501.

Other uses of barrel compost/CPP 502–507 include combining it with plant-based liquid manures (comfrey, stinging nettle, common horsetail 508) for the soil or directly on plants as a foliar feed with an additional anti-fungal effect provided (it is believed) by the anti-fungal properties of cow manure; spraying it on the root balls of young vines just before they are planted; helping the decomposition of matter in effluent ponds (farmers find it easier to add a few handfuls of barrel compost/CPP 502–507 than to fiddle with individual compost preparations); and as a seed bath. Plants grown in potting mixes come on earlier if doused with barrel compost/CPP 502–507. Barrel compost/CPP 502–507 can be added to heaps of manure or other material intended for composting before a proper compost heap is made.

ALEX PODOLINSKY'S PREPARED HORN MANURE 500 + 502–507 SPRAY

This soil spray was developed by Alex Podolinsky, Australia's most well-known biodynamic thinker and practitioner. Growers usually refer to it as '500P', but in this book, in order to be consistent in indicating by code number exactly which biodynamic preparation or preparations each spray or treatment contains, a slightly elongated description of it is used (see p.12).

Podolinsky was born in 1925 to a family of Russian-Ukranian aristocrats who had been forced to leave Russia during the 1917 Revolution. His mother introduced him to biodynamics through her contacts with farmers who had attended Steiner's 1924 *Agriculture* course or who had taken up Steiner's ideas, and who tutored Podolinsky at the Goetheanum in Dornach, Switzerland (see Appendix II). Podolinsky experienced what seems to have been a difficult education at public school in England but in 1945 fled to Australia after Churchill and Stalin agreed at Yalta to the setting up of the Russian Repatriation Commission. This threatened to send aristocratic families like Podolinsky's back to the Soviets and an inevitably brutal end.

Podolinsky felt that European biodynamic farmers' insistence on stirring their spray preparations by hand rather than mechanically meant that only small areas of cropland would ever become biodynamic. 'The world's population has increased four times in my lifetime and we can't feed everyone with small-scale backyard biodynamic farming,' Podolinsky argued. He reminded his critics that Steiner had accepted mechanical stirring as valid when giving the *Agriculture* course (see p.102). Podolinsky argued that mechanical stirring was capable of producing more consistent results than hand stirring, like deeper vortices in the water (although for recent advice to the contrary see Chapter 4). Podolinsky also felt Europe's biodynamic farmers had shown themselves to be no better qualitatively than their organic counterparts because their solid biodynamic preparations 500 + 502–506 appeared so dry as to be incapable of carrying the formative life forces lacking in contemporary agriculture. (For more on this debate see Chapter 2, storing compost preparations – dry or moist?)

In his adopted Australia, Podolinsky began refining his method for large-scale biodynamics. This, coupled with his interest in Pfeiffer's field spray concentrate (described above), led Podolinsky to develop a spray which he called the Prepared 500. Some Australians refer to it as 'powercow', while another name for it is prepared horn manure or 500P, although Podolinsky says to describe horn manure 500 as any kind of 'manure' is misguided.[22] Podolinsky's 'prepared 500' is made by combining the six compost preparations 502–507 with horn manure 500 and leaving the mixture in copper containers for one year. 'Prepared 500' is dynamized in water and sprayed on soil in the same way and in the same volume as horn manure 500.

Podolinsky's 'prepared 500' spray shares similar advantages to those offered by Maria Thun's barrel compost/CPP spray with the bonus of also allowing the horn manure 500 spray to be sprayed at the same time. Devotees of 'prepared 500' say it initiates biodynamic processes more strongly in the soil than Maria Thun's barrel compost 502–507 which (they claim) merely helps soil generate and digest organic matter (e.g. fallen leaves or recently ploughed-in cover crops). Podolinsky says his 'prepared 500' is not to be thought of as a long-term substitute for solid biodynamic compost 502–507, however.

French consultant François Bouchet suggested Podolinsky's 'prepared 500' risked creating an unwelcome opposition between growth forces and decomposition forces, resulting in excess vegetation, enhanced risk of fungal disease and reduced grape ripeness. Bouchet said horn manure 500 stimulates growth forces which in turn stimulate the *building up* of organic matter and enhanced root expression, while the compost preparations 502–507 exert forces which aid the *decomposition* of organic matter in compost (my italics). Bouchet claimed that while 'prepared 500' may build topsoil by 'animating' it, it does so only in such a way as to cause plant roots to remain in the topsoil, encouraging shallow (horizontal) rather than deep (vertical) rooting. While this may suit market gardeners growing mainly annual crops like vegetables, Bouchet saw it as problematic for perennial crops like vines reliant on a permanent, thus invariably deeper or at least more extensive, root system to express their sense of *terroir*. For this, classic horn manure 500 remained the best tool, Bouchet said (see Chapter 2, horn manure and the lime–silica polarity).

Yet what from Bouchet's European perspective appears a drawback may be a boon for those whose subsoils suffer salinity as a result of over-irrigation. Prepared 500's popularity in Australia may thus be a reflection of how important irrigation is to farmers there. Subsoils which exhibit both salinity and sodicity as a result of irrigation poison crops which root too deeply. Hence a soil spray like 'prepared 500' which builds topsoil without necessarily encouraging deeper rooting appears ideal under Australian conditions.

GREG WILLIS'S HORN CLAY SPRAY

Rudolf Steiner suggested clay was the substance which mediated between the two opposing force poles of lime/calcium and silica, the former being worked on by horn manure 500 and the inner planets (the moon, Mercury, Venus), and the latter by horn silica 501 and the outer planets (Mars, Jupiter, Saturn). Horn clay was developed in California in the 1990s by Greg Willis to act, he says, 'like the middle of the see-saw between the lime–silica polarities.' Horn clay is made by filling cow horns with clay-rich soil taken usually from the owner's farm. The clay is mixed to a slurry and the filled horns are usually buried for six months either over winter (winter/fall horn clay) as for horn manure 500 or over summer (summer horn clay) as for horn silica 501. Hugh Lovel (see also radionics: cosmic pipes and field broadcasters later in this chapter) claims the secret to mediating between the earthly horn manure 500 influence and the cosmic horn silica 501 influence is to bury horn clay for a full year so that both the winter/lime–calcium and summer/silica influences come to bear on the clay. Sceptics of horn clay, like Hugh Courtney, say that if both horn manure 500 and horn silica 501 are being used correctly then the clay will inherently be mediated, as Steiner made implicitly clear, making horn clay superfluous. Courtney says, 'horn manure [500] is all about metabolic [meaning animal digestive] processes which you don't get with horn clay. Clay is the mediator which should anyway come out of the rhythmic cow process. Horn clay will keep things on solely a material level [substances], without contributing the spiritual [forces] dimension which is what true biodynamic preparations are all about.'

FRANÇOIS BOUCHET'S URTICAE 500 SPRAY

François Bouchet developed a version of a spray originally conceived by German agronomist Volkmar Lust to stimulate vegetative growth (cell division) in newly planted vines, or vines whose growth has been weakened or blocked, especially vines suffering from fanleaf degeneration virus or phytoplasma diseases like Flavescence Dorée (grapevine yellows) and *bois noir*.[23] As its name suggests, Urticae 500 is made by combining stinging nettle (*Urtica dioica*) tea with horn manure 500. For one hectare Bouchet recommends 120 to 480 grams of horn manure 500 (up to four times the normal dose) and 2 kilos of fresh stinging nettles. The nettles are infused in boiled water, after which 1 kilo of willow (*Salix alba*) shoots and leaves can be added to prolong the infusion. This is left to cool (for up to twenty-four hours) before the horn manure 500 is added and the mixture is then dynamized for one hour. Urticae 500 is sprayed before flowering in spring when new shoots have grown to about 30 centimetres or five leaves, meaning on the upper or non-grape-forming part of shoots. Urticae 500 can be sprayed again on the same (non-fruit bearing) part of the vines after the embryonic grapes have formed in early summer.

HUGH COURTNEY'S SEQUENTIAL SPRAY TECHNIQUE

The sequential spray technique involves spraying all nine biodynamic preparations 500–508 in a thirty-six- to seventy-two-hour period when the sidereal moon stands in the leaf/water constellations Fishes, Crab or Scorpion. Hugh Courtney, of the Josephine Porter Institute for Applied Bio-Dynamics (see Chapter 2, horn silica 501 and the lime–silica polarity), developed sequential spraying to balance excess moisture in wet soils and increase moisture retention in dry ones.[24] The sequence begins when Maria Thun's barrel compost 502–507 is sprayed on the soil one evening. The following morning the same area is then sprayed with common horsetail 508 (as a liquid manure rather than as a decoction), pre-stirred for twenty minutes. Horn manure 500 is then sprayed later that afternoon followed by horn silica 501 the following morning. An abbreviated version, used in Australia, is to mix horn manure 500 with Maria Thun's barrel compost 502–507 before dynamizing (see Chapter 4) them together and spraying them on the soil one evening as the earth inhales, and spraying horn silica 501 the following morning as the earth exhales. This is said to have a balancing effect between the calcium forces exerted by the evening soil sprays and the silica forces exerted by the morning atmosphere spray (see Chapter 2, horn silica 501 and the lime–silica polarity, again).

PEST ASHING OR 'PEPPERING'

Pest ashing or peppering works by driving a vertebrate or invertebrate pest back to within its own natural geographical and population limits using its own burnt remains diluted in water as a deterrent. It provides an alternative to modern pesticides which tend to address the symptom rather than the cause of the pest problem. In biodynamics the element of water is seen as transmitting 'life forces' (see Chapter 4). With pest ashing the element of fire is used to transmit the negative of the animal's reproductive force, Steiner said.[25] Bringing the energy of the dead onto the land aims to repel the pest either directly by promoting strife within the species (fighting) or by interfering with reproduction (fecundity, male/female ratios) and thus adversely affecting population density and age structure (ratios of young to old).[26]

Pests for ashing should be collected from the farm rather than from neighbouring estates. With respect to vertebrate pests like rodents (mice, rats, rabbits and even marsupials such as possums), Steiner advised collecting the skin of the animal and burning it when the planet 'Venus is visible in the sky with the constellation of the Scorpion in the background. Venus has to be behind the sun.'[27] This is usually taken to mean behind the sun in terms of time because when Venus is physically behind the sun the latter's brightness prevents either Venus or the constellation behind Venus from being visible in the sky at all. As Steiner also referred to Venus as an 'evening star' this reinforces the idea of a temporal interpretation, because Venus is still visible after the sun has sunk below the horizon.

However, if you switch from the heliocentric model of our universe developed by Copernicus (1473–1543), in which everything revolves around the sun, to the geocentric model of 'dual centre' developed by Tycho Brahe (1546–1601), in which the planets still orbit the sun but the sun itself revolves around the earth, then what Steiner meant by 'behind the Sun' can be taken to mean 'behind the Sun sphere', meaning beyond the 'sphere' enclosed by the path of the sun around the earth. Steiner even referred to the Sun as a planet at the start of his 1924 *Agriculture* course, following Brahe's dual-centre model.[28] Under this model when Venus is in the Scorpion it is never directly (spatially) behind the sun and therefore both Venus and Scorpion are visible from earth. This does not happen every year but when it does the optimum period is usually between November and January.[29]

Making sprays from pest ashes is complicated.[30] Insects are incinerated by being placed in a pan or tin on a fire rather than on the fire directly. Tin cans – covered by a lid with holes in to allow smoke to escape without the ash also doing so – are burned first to remove any linings, paints or labels. A browny ash should remain, indicating that the temperature of whatever is being ashed had been kept relatively low rather than experiencing red heat. Peter Bacchus, who was born on a biodynamic dairy farm, ran commercial biodynamic greenhouses and served on New Zealand's BioDynamic Farming and Gardening Association for over twenty years, maintains that for peppering one should use the carbon ash rather than the salt ash. This is achieved by placing the ash in water, and stirring for a few seconds to dissolve the salt ash. This mixture is then strained through filter paper, and the water discarded. The filter paper is then dried and the ash collected once more. The ash is then crushed using a pestle and mortar for an hour.

One part of ash is diluted in nine parts alcohol and stirred for three minutes. This gives a D1 dilution. Successive dilutions to the same 1:9 ratio then take place to D3, a concentration which is said to be most suitable for storing. D3 potencies can be diluted to D6 or D8, the two potencies said to be the most effective, with 100 litres of D8 recommended for one hectare of vineyards.[31] The sprays must be prepared on the day they are to be used. Steiner said that as pest ashes work through the soil this is where they should be applied, rather than directly on crops. Peppers may work best on moist rather than dry soil. Bouchet suggests that a D8 solution of pest ash can be sprayed directly on the vines for pests of the vine canopy. Whether pest peppering works on all types of farm, only on biodynamic ones or best on biodynamic ones, is unclear.

WEEDS

Weeds are native plants that grow where farmers do not want them. Winegrowers with high aesthetic values seem especially intolerant of them. However, as Ehrenfried Pfeiffer pointed out, weeds can happily 'resist conditions which cultivated plants cannot resist, such as drought, acidity of soil, lack of humus, mineral deficiencies, as well as a one-sidedness of minerals, etc. They are witness of man's failure to master the soil, and they grow abundantly wherever man has "missed the train" – they only indicate our errors and Nature's corrections … Weeds are specialists. Having learned something in the battle for survival they will survive under circumstances where our cultivated plants, softened through centuries of protection and breeding, cannot stand up against Nature's caprices.'[32] See also weed manures, Chapter 6.

Weed ashing

One biodynamic technique to deal with weeds is weed ashing. It follows the same principles as pest ashing and is used to discourage weeds. Weed ashing may need to be used for four years on the same site to have a lasting effect. The reproductive organs of the weeds (the seeds, rhizomes, roots) from a single species are collected and burnt at full moon.[33] The ash can either be mixed with sand and sprinkled directly on the fields, or dynamized in water for one hour to make a spray or diluted homeopathically to a D8 dilution (see pest ashing or 'peppering', above).

Weeding by the moon

Weeds can be mown out or ploughed off in organic and biodynamic vineyards, but not weedkilled (unless organic-approved weedkillers based on pine resins are used). However, Maria Thun claims that mechanical weeding can be made more effective. Her research shows weeds germinate most strongly when the sidereal moon is in the fruit–seed/warmth constellation of Lion (a constellation which Thun says is especially favourable to seed formation) and germinate most weakly when the sidereal moon is in the root/earth constellation Goat.[34] One ploy is to plough under Lion to stimulate the greatest weed germination, then follow this by ploughing under Goat. Doing this in spring can knock weeds back for an entire season with any regrowth being just enough for protective ground cover but not so much as to compete excessively with vines, especially young ones. Weeds can also be squeezed out by sowing cover crops (see above), in effect replacing the weeds you don't want with ones you do.

WINTER TREE PASTE

Rudolf Steiner said everything above ground level has 'a particular tendency to life, a tendency to become permeated with etheric vitality.'[35, 36] This is why Steiner saw tree (or vine) trunks as elongated mounds of soil raised up above the ground. In his scenario tree bark represents the soil's surface and any branches, shoots or leaves growing from the trunk/bark resemble grass or other annual plants. To Steiner, the cambium or growing layer between the bark and tree trunk represented the subsoil or root system for those shoots and leaves.

Steiner was clear about the benefits to soil of biodynamic compost because the compost preparations 502–507 infuse compost piles with 'living forces, which are much more important to the [crop] plants than the material forces, the mere substance'.[37] Biodynamic winegrowers who want their vines to be manifestly 'permeated with etheric vitality', especially vines which are esca-affected or otherwise stressed, seek to give them a life-enhancing external stimulus between winter and spring. As

Applying winter tree/pruning paste to esca-affected Riesling vines

spraying the vines with compost is impractical, a tree paste (or pruning whitewash) can be applied instead to the vine's trunk and woody parts. This paste is to the vine trunk – an elongated mound of soil raised up above the ground in Steiner-speak – what compost is to soil.

Tree paste, or tree compost if you prefer, is made by combining what Proctor calls the three basic components of the soil – namely clay, sand and cow manure – with horn manure 500.[38] Proctor says to mix together one part each cow manure, silica sand (or diatomaceous earth), and potting clay (bentonite) or calcified sea algae (e.g. *Lithotamnium calcareum,* maerl). Mixing these with stirred horn manure 500 should produce a thin paste sloppy enough to apply to vine trunks with ease. Other choices for the liquid element include simple rainwater, Maria Thun's barrel compost 502–507 spray or teas, decoctions or liquid manures from common horsetail 508, stinging nettle, comfrey, seaweed and kelp teas. Propolis, whey or skimmed milk may also be added for their anti-fungal effect, both preventative and curative.[39]

Tree paste can be applied after leaf fall to strengthen the vine wood, after pruning to seal pruning wounds, or at bud burst to reduce excoriose (phomopsis). The paste can be applied with a thin whitewash brush by hand on a small number of fruit trees or vines, or with a sprayer on larger numbers. If it is to be sprayed, Proctor advises making it with bentonite and diatomaceous earth rather than with potting clay or silica sand so it is easier to spray, and to use a coarse nozzle. A centrifugal pump is less likely to be damaged by the nature of the materials than a diaphragm pump. He advises applying tree paste in winter, after pruning but before bud burst, and under a descending moon.

Masson suggests tree paste be used on pruning wounds on plants where diseases of the wood are a concern, sealing over crevices in which pests may overwinter, and sealing and healing wounds created by pruning. Removing any dead, dried or flaky bark, moss and lichen first is very time consuming but means the cambium is more likely to be exposed to the benefits of the tree paste.

PRUNING WASH

François Bouchet describes a pruning wash similar to one used by fruit growers.[40] This is applied in autumn after leaf fall on the vine wood or on the soil, then again at the end of winter after pruning. The bacteria contained in the pruning wash clean vine wood of pathogens. For one hectare of vines, 240 grams of Maria Thun's barrel compost 502–507 is mixed with between 1 and 2 litres of whey from cow, goat or ewe's milk. In addition, bentonite or montmorillionite clay is premixed into a lumpy paste by diluting it in water, and this is then added to the manure/whey mixture at a dose of 1 per cent. The clay simply helps stop spray nozzles from blocking. Before use, the barrel compost 502–507, whey and clay mixture are diluted together in water and dynamized for twenty minutes. The dynamized spray should be used within forty-eight hours of preparation. Bouchet says one should always leave a three-week gap between applying pruning wash and spraying horn manure 500, so as not to create an opposition between the forces in the horn manure 500 (building up) and those carried by the six compost preparations 502–507 (decomposition) via Maria Thun's barrel compost.[41]

Pierre Masson's pruning wash is made by soaking 15 litres of fresh cow manure in a porous sack in 100 litres of rainwater to make a manure tea.[42] Then, in another recipient, 15 kilos of dry clay – preferably kaolin or potting clay (which Masson says is less easily washed off by rain than montmorillionite or green clay) – is mixed in 100 litres of rainwater to make what should look like a light clay soup. This will help the paste to stick. The clay soup and manure tea are then mixed together. Masson suggests adding 20 litres of whey too, if it is available. The mixture can be topped up with 80 litres of water, or water mixed with either horn manure 500, prepared horn manure 500 + 502–507, Maria Thun's barrel compost spray 502–507 or common horsetail 508 (decoction).

RADIONICS: COSMIC PIPES AND FIELD BROADCASTERS

Field broadcasters, also referred to as 'cosmic pipes', are a modern invention. They are made from a (PVC) pipe which is placed vertically in the ground and is usually 3 metres high. Inside the pipe is a dual crystal coil circuit. This is said to work off streams of energy flowing upward from within the earth and others flowing downward from the celestial sphere. Inside the pipe are glass phials with the biodynamic preparations in tablet form. Via separate reagent wells and coils for both upward and downward streaming forces the pipe is said to transmit the effect of the biodynamic preparations across the farm and into the earth and the atmosphere in a form of radionics.

Manufacturers of field broadcasters, such as Hugh Lovel in Australia, say they are cost effective because they broadcast the effect of every biodynamic preparation at a constant, albeit low, frequency level, whereas the effect of applications of either the three biodynamic field sprays or biodynamic compost is only intermittent. The disadvantage, of course, is that field-broadcaster technology was never described by Steiner in his 1924 *Agriculture* course, is not part of traditional biodynamic practice, is not recognized by Demeter as a substitute for applying the biodynamic preparations in spray and compost form to farmland, and is not proven to work.

Hugh Courtney says, 'field broadcasters may tell the farm it'll be getting some biodynamics, but without actually giving the farm any biodynamic preparations. It's like getting to know someone by seeing them via a photo or on DVD rather than in the flesh.' In other words if biodynamics was as easy as simply sticking a prep-filled pipe in the ground, then perhaps everyone would be doing it.

6

PLANT TEAS, DECOCTIONS, LIQUID MANURES, OILS AND OTHER ALTERNATIVE TREATMENTS

The use of teas, liquid manures, extracts and essential oils based either on wild plants or composted matter is widespread in biodynamic viticulture and becoming more common in general wine growing too. This results from a greater understanding of the beneficial role the micro-organisms (yeasts, fungi, bacteria) which these sprays contain play in keeping vines healthy by colonizing both rhizosphere (root zone) and phyllosphere (leaf surface). Plant- and manure-based sprays can help reduce or eliminate reliance on anti-fungal treatments based on sulphur (for oidium) and copper (downy mildew) which, although approved in organics as contact sprays, can have adverse effects on soil and vineyard biota (see p.169).

Such plant- and manure-based sprays are generally easy, quick, cost-effective and even fun to prepare since the raw materials can often be gathered wild locally or grown on the estate. This suits winegrowers working towards the biodynamic goal of turning the vineyard into a self-sustaining living organism. Locally gathered plants adapted to local conditions are more effective at preventing or even curing diseases or imbalances specific to that locality.[1] Commercially produced bacterial or fungal compost cultures are now widely available but – apart from being expensive to buy – may adversely affect the balance of existing local microfauna and microflora whose main stimulae should be the biodynamic horn manure 500 and compost preparations 502–507.[2]

The quality of the water used for herbal sprays is vitally important, as it is for all biodynamic sprays (see Chapter 4, the water). Liquid manures and teas specifically aimed at influencing vine fertility are commonly sprayed in the evening or during the period of the descending moon, when vines, the soil and vine roots are all breathing in, allowing the effect of the spray to be maximized.

François Bouchet said the beneficial effects that plant-based sprays have is considerably strengthened by dynamizing them with biodynamic horn manure 500.[3] Bouchet pointed out that plant-based sprays in general, and liquid manures in particular, should ideally always be dynamized (see Chapter 4).

Different plants can be combined in a single spray, an example being stinging nettle plus either chamomile or yarrow tea. Suggestions given in biodynamic literature regarding how plant and compost sprays are best used do vary, reflecting the role local conditions and personal preferences play. For instance, no consensus exists in France on whether stinging nettle and common horsetail 508 should be combined in a single treatment to reduce downy mildew pressure. Growers for whom these two plants form the backbone of their plant-based spray regime may chose to use both but alternately with copper-based anti-fungal sprays: stinging nettle tea at lunar apogee to accentuate the beneficial effect nettle has on photosynthesis, and common horsetail 508 at lunar perigee to increase crop resistance.

The incorporation of whey, milk and skimmed milk into sulphur treatments to make their anti-fungal effect against powdery mildew more effective is increasing. See also their use in winter tree paste and pruning wash, pp.120 and 122.

PLANT TEAS AND INFUSIONS

Wild plants have been used in medicine for thousands of years. Note, however, that whereas plant-based remedies for treating humans and animals are often made from the plant roots (e.g. dandelion, valerian) it is generally the flowers, shoots, leaves or even the bark which are used for treating crops (or for treating compost with the six biodynamic compost preparations 502–507).

The seven plants Rudolf Steiner directed be used as biodynamic preparations – yarrow, chamomile, stinging nettle, oak bark, dandelion, valerian and common horsetail – are all popular choices for use as vineyard sprays although stinging nettle – 'the greatest benefactor of plant growth' said Steiner (see p.54) – is perhaps the most commonly used of all. Note that the five plants used for the solid compost preparations 502–506 are utilized in their wild state – as flowers, stems and leaves or bark – rather than having first been transformed into actual biodynamic preparations. It may also be worth noting that the solid compost preparations 502–506 are usually only diluted in water individually when preparing a bath for seeds to soak in before they are sown. Certain compost preparations seem to have affinities with certain seeds.[4] Seed soaking is rarely used by winegrowers as vines are propagated vegetatively from cuttings rather than sexually from seeds, although some biodynamic growers are trying to establish vines derived from sexual reproduction. It is, however, common practice to add a set of biodynamic compost preparations 502–507 to liquid manure sprays (see below).

While on the biological level plant-based sprays stimulate growth by providing nutrients or substances for vines, biodynamicists also say plants have affinities to certain celestial bodies, and so carry beneficial cosmic and earthly forces into the vineyard, too. Winegrowers following celestial rhythms (see Chapter 7) try to apply plant-based sprays when the sidereal moon stands in the fruit–seed/warmth constellations of Ram, Lion and Archer, these having the most affinity with wine grapes.

Plant teas are made by pouring freshly boiled water over fresh or dried flowers and herbs and letting the mixture infuse before it is drained. Maceration times tend to be short, usually less than twenty minutes or the time it takes the water to cool. This is because the subtle, light, airy qualities that flowers like chamomile, dandelion, valerian and yarrow possess are said to be inhibited by excess heat or overly extractive macerations, although woodier aromatic flowering plants like rosemary may be left to infuse for up to twenty-four hours. Decoctions are more extractive and are discussed on page 134. For all teas fresh or freshly dried plant material is preferred. Around 10 grams of fresh plant material is enough for one hectare of vines when using only the flower heads, while around 100 grams per hectare is recommended for plants like comfrey or stinging nettle whose shoots and leaves are used. Commonly a small volume of concentrate is made which can subsequently be diluted later in more water (e.g. 35 to 200 litres per hectare) or combined with the usual anti-fungal vineyard sprays (e.g. copper, sulphur).

Pierre Masson's observation that plant-based sprays are most effective on land regularly treated with the biodynamic preparations, farmed to high agronomic standards and with the greatest possible diversity of animals and plants, is worth noting.[5]

Chamomile tea

Chamomile is a popular tea amongst biodynamic winegrowers. It is made from the flowers of the same strain of chamomile used in the biodynamic compost preparation 503, and picked in the same way (see Chapter 2). Chamomile is rich in sulphur, so blending its tea with the sulphur sprays used against oidium is said to make them more effective. Chamomile's calcium and potash content also helps stimulate healing processes in vine shoots and leaves damaged by hail or pruning.

Chamomile helps clear blockages. Its 'fine roots break up compacted soil, giving structure to the topsoil', say von Wistinghausen *et al.*, adding that in humans chamomile relieves blocked digestion by bringing a 'dissolving and healing process to the mucous membranes in particular. It can have demulcent and healing properties where inflammatory changes or hardening occur in the gastrointestinal tract.'[6] In vines chamomile stimulates sap movement during potentially stressful weather extremes of heat, cold, rain or drought, especially if sprayed during the descending moon. Thun says chamomile tea is best sprayed on vines at sunrise.[7] See also valerian tea, below.

Dandelion tea

Dandelion roots are used in remedies to treat humans, notably to relieve skin ailments. For crops, however, the flowers are used and are picked in the same way as for the dandelion compost preparation 506 (see Chapter 2). Dandelion contains many nutrients including calcium, copper, iron, magnesium, potassium and silicon. Dandelion's silica is especially useful in strengthening and tightening vine leaves against fungal parasites in wet years or when full moon and lunar perigee coincide.[8] It is recommended as a preventative spray at the start of the vine's vegetative cycle after the first two or three leaves have appeared. It can be used alone or combined with small quantities of copper and sulphur vineyard sprays in spring.[9]

Stinging nettle tea

The most frequently used and arguably most effective tea used on biodynamic vineyards is made from the shoots and leaves of stinging nettles. These can be picked in spring or until midsummer, as for the stinging nettle compost preparation 504 (see Chapter 2).[10] Stinging nettle's multiple beneficial effects include increasing resistance to pests and diseases, aiding photosynthesis and stimulating growth by providing a long list of trace elements including iron, potassium, sodium, sulphur, calcium, chromium, selenium, silicon, cobalt, zinc, magnesium and manganese. Countless French biodynamic winegrowers say stinging nettle alleviates chlorosis (leaf yellowing) caused when iron and manganese become blocked on alkaline soils.

Stinging nettle tea is often combined with copper and sulphur sprays to minimize the hardening effect copper has on vine leaves, meaning lower doses of it can be used, as long as the pH of the mixture is neutral or slightly acidic.[11] Stinging nettle's role in promoting sap flow in hot weather helps deter aphids like red spider mites which colonize parched leaves and vine wood. For this a cold nettle extract is sprayed in summer just before lunar apogee, a period during which the moon is said anyway to evoke a 'summer mood'. The cold extract is made by following Thun's recommendation of soaking fresh stinging nettles, which can be in flower but must not have gone to seed, in cold or lukewarm water for twenty-four hours.[12] She says to spray it twice in a matter of a few hours but this is easier for small allotments than most vineyards. Sattler and Wistinghausen suggest stinging nettle can be combined with horn silica 501 sprays.[13]

As a soil spray, stinging nettle suppresses the upward movement of fungal disease spores, especially in the days leading up to full moon[14] or when the sidereal moon stands in the leaf/water constellations Scorpion, Fishes and Crab.[15] In India, Proctor found that blending 5 per cent fresh cow's urine in nettle tea (made there from *U. parviflora* rather than *U. dioica*) is effective against chewing and sucking insects.[16]

Courtney says that diluting the solid stinging nettle biodynamic compost preparation 504 in water produces a spray which, when applied separately or used as a watering-in agent around plants, strongly stimulates leaf growth and can be of some benefit in preventing frost damage.[17]

Note: Stinging nettle may act as a host plant for *Hyalesthes obsoletus*, an insect vector of an incurable vine wood disease, *bois noir*.

Valerian tea

Valerian tea is made either by infusing valerian flowers in freshly boiled water, or as a concentrate in exactly the same way as the biodynamic compost preparation 507 (see Chapter 2). Just as humans take valerian tea as a sedative, winegrowers can spray vines with valerian tea as a calmative after the stress of pruning. It can also be used to destress vines hit by hail, sprayed either on its own or added to any anti-downy mildew treatments if these are also being used, perhaps combined with a few drops of arnica tincture.[18] Valerian warmth forces may also stimulate grape ripening and flavour development if sprayed in the weeks before harvest (possibly in combination with horn silica 501). Combining valerian tea with the first horn manure 500 or prepared horn manure 500 when this is being sprayed on the soil around bud break may also mitigate against frost. In this case add the valerian liquid at the very start of dynamization.[19]

Valerian owes its warming, soothing effect to its phosphorus content which activates warmth processes. For vineyards likely to be affected by frost valerian tea acts as a protective heat blanket, the idea being that even though the sun may shine on frosty ground its rays become especially warming in valerian's presence. However, Nicolas Joly says although it may be comforting to think of spraying valerian as a pre-emptive anti-frost measure, essential oils (see below) from thyme or clove may be preferable. 'It would be better to think of valerian's heat as enhancing flavours in the grapes, especially for vines whose maturation has been slowed by a cold spring,' he says. This follows Thun's idea that valerian sprayed on the soil in late afternoon at midsummer when the sidereal moon stands in the flower/air constellations Twins, Scales and Waterman will stimulate vine bud development for the following year;[20] although overuse at this time may cause existing vine flowers to wilt.[21] If valerian tea has been sprayed on the vines between bud burst and flowering Bouchet suggests spraying chamomile tea after flowering to counter any potential tendency to cause wilting which the valerian might have stimulated.[22]

Yarrow tea

Yarrow tea is made from yarrow flowers which are picked as for the compost preparation 502 (see Chapter 2). The tea's main role is to make sulphur sprays against oidium more effective by activating potassium or light processes in the vines, revitalizing them in the same way that people who are feeling run down drink yarrow tea in the morning as a pick me up. Some growers claim yarrow tea can make the subsequent wine less susceptible to oxidation.[23] Yarrow's other role is to help cane ripening if sprayed around *véraison*. Adding some stinging nettles to macerate with the yarrow flowers produces a tea capable of regulating both fungal diseases and insects.[24]

Other plant-based treatments

Birch leaf tea is used by some biodynamic wine growers in Austria to counter downy mildew. Borage tea (*Borago officinalis*) works in a similar way to stinging nettle tea but is richer in silica. Meadowsweet (*Spirea ulmaria*) works in a similar way to willow and is used by winegrowers in the Jura, but the beneficial effect of its salicylic acid is nullified by heating the water beyond 80°C.[25] Tansy tea (*Tanacetum vulgare*) or scented fern has a slightly resinous aroma which is said to repel parasites and may have a fungicidal effect against downy mildew. Rosemary (*Rosemarinus officinalis*), sage (*Salvia officinalis*) and garlic (*Allium sativum*) have antiseptic and repellent effects, and are used against grape berry moths by winegrowers in Mediterranean France. Adding mint to teas helps make the teas more durable, especially in hot climates, via mint's antioxidant properties. Mint can be added as leaves or as an essential oil. An extract (brand name Stifénia®) made from fenugreek flour (*Trigonella foenum-graecum*) seems to improve vine resistance against powdery mildew before flowering.

COMPOST TEAS

Compost teas are made by aerating finished compost in water. Aeration activates beneficial micro-organisms and helps them multiply. Compost teas improve soil biology when sprayed on the soil, and as foliar sprays they aid disease suppression and nutrient uptake in crops. Over the last decade the Oregon-based microbiologist Dr Elaine Ingham has become a leader in the field of aerated compost teas. Ingham has established a network of

Soil Food Web (www.soilfoodweb.com) laboratories in Europe, North America, South Africa and Australasia. These labs allow farmers to analyse the microbial life on their crops, in their soils, in their compost and in their compost teas.

Compost teas need the right balance of nutrients and the right range and populations of beneficial micro-organisms. The base for the kind of fungal-dominated compost teas which are favoured in winegrowing could be made from well aged (six to eighteen months' decomposition) woody compost (from carbon-rich woodchips and chipped vine prunings or woody weeds like gorse and broom), Peter Proctor's cow pat pit preparation 502–507, hay steeped in liquid manure slurry, seaweed tea, vermicasts and vermiliquids, molasses, basalt and lime.

Once mature, the fermented compost is mixed in good quality water (ideally very slightly acidic, degassed of any chlorine) and then aerated. Aeration can last up to twenty-four hours and is necessary because most beneficial bacteria, fungi, protozoa and nematodes are aerobic. Microbes present in compost teas must be alive when they are applied to crops or soil so they can produce exudates that help them stick to the crops, even in heavy rain. Spraying compost teas as a fine mist and at low pressure protects the micro-organisms' cell structures. Rates are commonly 1 kilo of compost aerated in 100 litres per hectare for a soil drench, and double this for a foliar application (depending on vine canopy height).

Soil drenches are effective during periods when high levels of soil microbes need to be encouraged: in spring, before flowering, after *véraison* and after harvest. For a foliar application around 70 per cent of the leaf surface needs to be colonized by the beneficial micro-organisms for them to be effective.

Ingham says the beneficial micro-organisms in compost teas have several roles. They can induce what is called systemic-induced resistance in crops.[26] Microbes in the compost tea surround the plant surface, but because plants cannot differentiate beneficial micro-organisms from non-beneficial ones an immune response is triggered, making plants more resistant.

Fifty per cent of the carbohydrates produced in crop leaves end up in the root system. It makes sense, therefore, for any food that plants produce as exudates from their roots or leaves which is intended specifically for soil bacteria and fungi is accessed only by beneficial disease-suppressive organisms rather than disease-causing ones. Once safely *in situ* beneficial organisms perform what is called niche occupation, meaning they take the space disease-causing organisms would otherwise occupy, and they take their food too. They can even predate disease-causing organisms (protozoa feed on bacterial disease organisms, for example) and produce compounds that inhibit their growth. Beneficial bacteria and fungi play the key roles in protecting crops above ground, while below ground – where bacteria and fungi are key – protozoa and nematodes also play vital roles.

Winegrowers increasingly say they want fungally dominated vineyard soils, commonly a ratio of between 2.5:1 fungae to bacteria. The vine is a woody plant and woody plants need fungal-dominated 'forest-floor' soils dominated by mycorrhiza. Soils too high in bacteria mean vine roots may lack the mycorrhiza, allowing them access to the soil system or the soil food web as Dr Ingham calls it. The greater diversity of that mycorrhizal fungi, the greater diversity of nutrients they can access, meaning plants feed more efficiently (saving on fertilizer). Beneficial microbes also make the soil more airy and freer draining. Weedkillers and synthetic fertilizers inhibit mycorrhizal fungi and create a more bacterial system.

Compost teas are not mixed with regular vineyard sprays containing copper or sulphur; in fact, the aim is for the compost teas to render the use of copper or sulphur redundant. However, compost teas can be used in rotation with sulphur and copper sprays in order to maintain microbial diversity in the vine's leaf canopy. Winegrowers spraying eradicant or broad-spectrum fungicides eliminate rot-causing organisms by sterilizing a horizontal strip along the vine canopy where the grape bunches grow. However, because disease organisms live throughout the leaf canopy, both above and below the sprayed zone, they can quickly recolonize the sprayed area – for example, when washed down there by the next rain. The disease organisms then find the grape bunches provide them with a perfectly sterile site upon which to grow and quickly reproduce. However, if compost teas are also being used then no single fungal organism, especially the aggressive disease-bearing ones, can easily become dominant, which reduces disease pressure.

All the nutrients that beneficial bacteria need should already be present if the compost being used for the tea has been well prepared.

If not, food sources like molasses, complex sugars, fulvic acid and vegetable juice can be added to the compost tea. Food sources for fungi include kelp/seaweed extracts, rock dust and fish oil/emulsion, or humic acids. In the United States because of fears of *E. coli* compost teas (even aerated ones in which *E. coli* cannot survive) made from or using raw manures can only be applied to vineyards no later than ninety days before the grapes are harvested.

Compost teas should also be used in conjunction with regular teas when the latter are being used as foliar feeds. Crops take in more nutrients (from foliar feeds) when their leaves are covered in bacteria and fungi (from compost teas) because these organisms respire carbon dioxide. Carbon dioxide around the leaf surface causes the leaf stomates to open quicker and to stay open for longer than usual. This then helps plants assimilate any nutrient material contained in the foliar feed more easily, potentially reducing the frequency, strength and cost of foliar feed applications.

PLANT DECOCTIONS

Decoctions are made by placing shoots, leaves, flowers or bark of the chosen plant in cold water, bringing the mixture to the boil and if necessary then leaving it to simmer before filtering off the concentrate and diluting this to between 1:5 to 1:20 before applying it to the vineyard. Decoctions involve longer, hotter and more extractive macerations than for teas and infusions, and are suited to extract the calcium from oak bark, the salicin from willow or the silica from common horsetail.

Oak bark decoction

The oak bark decoction is derived from the bark of the same strain(s) of oak gathered in the same way as for the biodynamic compost preparation 505 (see Chapter 2). Oak bark decoction can be used to complement the common horsetail 508 decoction, reining in excess 'moon forces' which make vines grow vigorously enough to attract fungal diseases like rot and mildew, especially around full moon. Oak bark has high levels of both calcium (about 75 per cent of the ash is CaO) on the one hand, and tannin on the other – the reason why 'tanbark' was widely used in tanneries.[27] Tannic acid has insecticidal properties while calcium protects against fungal growth. The oak bark decoction is seen by some winegrowers as creating grapes with hardier, more 'tannic' skins, tightening them to make them more resilient against fungal attack. Overuse of this decoction, however, may cause a severe blockage in plant growth.[28]

To make enough decoction for one hectare, Masson advises placing 50 grams of oak bark (which can be unground chunks) in 3.5 litres of cold water which is brought to the boil and left boiling for fifteen to twenty minutes.[29] Others macerate or simmer the oak bark in water for up to an hour. The concentrate is filtered off and then diluted in 35 litres of water.

Willow decoction

Willow decoction is made from the shoots, leaves or bark of the white, common or European willow tree (*Salix alba*). The willow grows in humid riverbank conditions yet never suffers from mildew, so its main role is anti-fungal (downy mildew, powdery mildew, botrytis), a role it shares with both the common horsetail (*Equisetum arvense*) 508 spray preparation and oak bark decotion – but it is easier to harvest than the latter because more than the bark can be used.

Willow's active ingredient is salicin. In the human body salicin converts into salicylic acid, the forerunner of aspirin but without the latter's stomach-irritant effect. Salicylic acid levels are highest in willow in spring which is when plant material should be collected.[30]

Willow is popular amongst winegrowers with grape varieties highly susceptible to downy mildew like Grenache Noir.

Plant material from both willow and stinging nettle is often combined when making vineyard teas and liquid manures, although willow seems more adapted to decoctions (simmering) than stinging nettle. See also Chapter 5, François Bouchet's Urticae 500 spray.

LIQUID MANURES

Liquid manures are made by leaving plant material to decompose and ferment in water over a period of days, weeks or even months. Typically winegrowers fill old barrels with around 100 to 200 litres of rainwater, plunge as many armfuls as they can of whatever it is they are macerating in the water (5 to 10 kilos is normal), then leave the mixture to warm in the sun. This is called passive brewing. Wrapping the plant material or animal

*These manure pouches have been macerated in water to make liquid
manure destined for this Chilean winery's vineyard irrigation system*

manure (see photo) in cheesecloth or using wooden batons will keep it
submerged. A set of biodynamic compost preparations 502–507 can be
added prior to the soaking period. For this, the five solid biodynamic
preparations 502–506 can be placed in balls of compost wrapped in a
cheesecloth and either dropped in directly or suspended by string from a
crossbar. The valerian 507 liquid can be dripped in directly. The growth of
beneficial micro-organisms can be encouraged in liquid manures by adding
a similar range of foods to those used in compost teas (see above), then
aerating to encourage the microbes to reproduce whilst making the liquid
manure smell nicer. Bouchet says liquid manures smell strongly even when
diluted 1:10 in fresh water for use on the soil or 1:20 for use on vine leaves,
and must be used with care on vineyards after flowering if the taste of the
wine is not to be impaired.[31] Excessive use of liquid manures, especially
those which have not been aerated, is discouraged in winegrowing.

Comfrey liquid manure

Comfrey (*Symphytum officinale*) is a popular liquid manure in organic
farming and home gardening. Like stinging nettle, comfrey is a rich source
of nutrients and trace elements: iron, phosphorus, calcium, potassium,
sodium, manganese, chromium and selenium. Comfrey is especially good

for fruiting and seed-filling crops because it stimulates potassium processes, aiding photosynthesis and reinforcing plant self-defence mechanisms, e.g. against downy mildew in viticulture.[32] It de-stresses vines after hail and, like seaweed, can aid boron deficiency at flowering. The shoots and leaves are picked at the beginning of flowering. Comfrey can also be used as a tea.

Fern liquid manure

Fern (*Polystichium filix-mas*) makes a resinous-scented liquid manure which French consultant Michel Barbaud claims can discourage attacks from grape berry moths and the leaf hopper vector of the Flavescence Dorée (grapevine yellows) phytoplasma disease, especially when used in conjunction with essential oils (see below). Barbaud's method is to soak 15 kilos of dried fern in 1,000 litres of cold water which is left outside to warm under the sun for around four days if it is to be sprayed on the vines, and up to ten days if it is to be sprayed on the soil. For each hectolitre of fern liquid manure Barbaud says to add 50 millilitres of essential oil of fennel (a well-known insect repellent for home gardeners) if spraying before *véraison*, or 20 millilitres of essential oil of garlic if spraying after *véraison*, although this latter strategy may, depending on picking date, impart a negative taste to the wine.

Seaweed and kelp liquid manures

Seaweed extracts are popular products in health food stores for their magnesium content. In the late 1960s, France's first organic wine and farm consultants, Raoul Lemaire and Jean Boucher, recommended that powdered, calcified sea algae (*Lithotamnium calcareum*) be used as a magnesium-rich soil amendment in vineyards at a rate of 80 to 100 kilos per hectare.[33] The Lemaire-Boucher method, as it became known, maintains magnesium is to vine sap what iron is to our blood.[34] (Levels of magnesium and calcium must be balanced in soil for healthy crop growth). Seaweed is also super-rich in those several dozen trace elements and salts which can be lacking in soils which were never part of the sea floor. New Zealand winegrowers who use seaweed do so because vines there can lack boron and molybdenum at flowering. They follow Peter Proctor's instruction to half fill a non-corrosive container (like a 225-litre barrel) with fresh, chopped seaweed or kelp, or with a 12 kilo measure of powdered or ground and dried seaweed.[35] The drum is topped up with rainwater and left to ferment outside, with aeration via stirring every two to three days. It should be ready

Angus Thomson finds liquid manures from seaweed, broom, ragwort, willow and scotch thistle provide trace elements for his Urlar Estate on the potentially acidic silt loams of the Gladstone sub-region of the Wairarapa, New Zealand. Thomson mixes the liquid manures with Peter Proctor's cow pat pit spray (CPP) 502–507 and horn manure 500 and dynamizes them in a 600-litre flowform.

in around two months, when the liquid will be sweet smelling and a clear brown colour. The concentrate should be strained and used at a rate of 6 to 11 litres diluted in 110 litres of water for one hectare. It should be sprayed at least three times a year as a tonic, a source of trace elements and a stimulus to the potash processes.

Stinging nettle liquid manure

This is the most commonly used liquid manure in vineyards. It is made from the stems and leaves of stinging nettles which are cut before or at the beginning of flowering.[36] Its effect is similar to but more powerful than the fresh tea version of stinging nettle described above. Stinging nettle liquid manure is generally sprayed on the soil or on vines as a pick-me-up after early spring frost and against early season chlorosis. It combines especially well with horn manure 500 when used as a soil spray. As a mite repellent, the cold extract or tea form of stinging nettle is more effective than the liquid manure. Courtney advises using this liquid manure with care 'as

overuse, say, more than twice in a growing season, can cause the plants …
to be more subject to fungus attacks, in the same way that happens when
one uses too much raw, uncomposted manure'.[37]

Weed manures

Weeds out-compete cultivated crop plants and as a result contain vital
nutrients often missing from soils – nettles and dandelions are high in
calcium, iron and magnesium, while thistles are high in phosphorus as
well as trace elements like zinc and manganese. Balance can be restored by
macerating weeds in water to make a concentrate, then spraying this on the
soil. The extract can be made from the branches, leaves, flowers and roots
of whichever weed predominates. Ferments made from weeds which spread
through rhizomes and stolons are supposed to be especially beneficial.[38]
Weed manures should be made from just one weed type at a time, as mixing
species may interfere with the breakdown process.[39]

ESSENTIAL OILS

The use of essential oils of the type used in aromatherapy is increasing
amongst winegrowers, especially in France. The volatile aromatic
compounds in essential oils can play a role in repulsing insect pests and
stimulating crop resistance to fungal disease organisms. The aromatic
compounds are expressed either by pressing (cold extract) or by distillation
in steam. Examples include eucalyptus, used early in the season against
downy mildew; palmarosa against fungal disease organisms; fennel and
lavendin [sic] to repulse various insect vectors; garlic and fennel against the
grape berry moth. Essential oils of thyme or clove can even be used against
frost (see valerian tea, above). Sage, citronella, Sylvester pine, grapefruit and
eucalyptus are other bases for essential oils used in viticulture.[40]

Only tiny volumes of raw material are needed, one or two heads of
garlic or a handful of rosemary stems per hectare of vines, for example.
Cold extracts are made by crushing and macerating the raw material in
olive oil for a couple of days and then decanting the oil into a bottle
containing a fatty substance to emulsify the active ingredient so that it
mixes in water ready for application. Cold extracts are said to retain a
greater range of active elements than when either steam or hot water
are used.

Essential oil sprays can be used sparingly around the perimeters of vineyards (the oils are very costly to buy), or on just a few vine rows per plot, and infrequently, as deterrents. Most growers use them only up until *véraison* so as not to affect the taste of the wine. They are photosensitive (light sensitive), and so must be used in the evening.

7

WORKING TO CELESTIAL RHYTHMS

As well as being the oldest organic movement, biodynamics was modern farming's first attempt to take active account of the movements of and forces exerted by the moon and other planets, and by the sun and other stars, when timing agricultural work. Our planet was formed by and forms part of a wider universe. Biodynamic methodology assumes that life originates from the whole universe rather than solely from what the earth provides.

From the very beginning of his 1924 *Agriculture* course Rudolf Steiner stressed the important influence celestial rhythms have on humans, plants and animals. But he warned that human life's 'emancipation from the cosmos is almost total. It is less so for animals, and the plants are to a great extent still embedded in and dependent on what is occurring in their earthly surroundings. This is why it is impossible to understand plant life without taking into account the fact that everything [which happens] on Earth is actually only a reflection of what is taking place in the cosmos.'[1] We humans really do seem to have lost touch with what is going on above our heads. More than half the world's population now lives in cities where light pollution means only the very brightest objects in the sky remain visible even on the clearest of nights. The most obvious celestial rhythms affecting us are being usurped in other ways, too. Our twenty-four-hour culture nullifies the day–night rhythm, throwing the body's 'circadian clock' (the inbuilt mechanism that regulates waking and sleeping) out of balance, leading to dietary problems and the kind of cognitive decline in shift workers more commonly associated with acute premature ageing. Our sense of seasonality is being challenged by hydroponically grown and air-freighted food which gives us fresh strawberries and tomatoes an unnatural four seasons out of four.

Although science accepts that the moon's position relative to the earth can influence the earth's shape, weather, atmospheric pressure and tidal movements it disputes the effect, for example, the full moon has on when women give birth; and scientists view as simply fantastical some of Steiner's more esoteric philosophies – for example, that the inner and outer planets have a relationship with calcium and silica respectively (see p.156).

THE BIODYNAMIC FARM ORGANISM AND THE COSMOS

Nevertheless, Steiner viewed each farm as an organism, the organs of which are the soil, the crops, the animals and the farmer – and by extension each farm organism was to be considered part of the whole cosmos, and not as just earthly or earthbound. Steiner said, 'The plant is like a butterfly chained to the ground; the butterfly is like a plant liberated by the cosmos.' Ancient astronomers such as Hesiod had produced agricultural calendars linked to celestial movements in the late eighth century BC. In the sixteenth century German mathematician and astronomer Johannes Kepler (1571–1630), who discovered that the movement of the planets around the sun was elliptical rather than circular, produced almanacs that claimed to predict the nature of the harvest for farms and even vineyards. Steiner himself produced a cosmic *Kalendar* in 1912–13, but this was lunar rather than planetary, and concerned the position of the moon against the background of the twelve main zodiacal star constellations lying behind the ecliptic. This is the path the moon, the sun and the planets all follow across the celestial sphere as defined by the earth's orbit around the sun. It is called the ecliptic because it is the point at which eclipses can occur if one celestial body 'meets' or blocks another from view.

The precession of the equinoxes

The signs of the tropical zodiac given by astrologers date from ancient Greece. At that time in the northern hemisphere the sun stood in Ram on 21 March, at the vernal point meaning the point in the zodiac where the sun stands when it is exactly overhead on the equator. This is when day and night are of equal length over the whole earth, and is the moment of the spring equinox. However, the sun moves backwards through the astronomical zodiac by one degree every 71.5 years (thus it takes the sun a 'Platonic year' of 25,920 years to make the complete 360-degree circuit). This is called the precession of the equinoxes. Its backward movement means that the sun now enters Fishes on 11 March, and therefore this is the constellation the sun stands in on 21 March, not Ram (or Aries, which is what the astrology column in your daily newspaper appears to suggest). The sun now enters Ram on 18 April. By 2375, the sun will stand in Waterman at spring equinox.

Steiner's *Kalendar* avoided doing what the zodiac of popular astrology does, which is to divide this band into twelve equal 'houses' or signs of 30 degrees for each, whose names are normally given in Latin. Instead Steiner focused on the actual astronomical, physical, position in the sky of the moon against the background of these twelve main constellations, all of which are of unequal size. In astrology the signs are referred to using Latin names (Aries, Taurus, Gemini, Cancer, Leo, Virgo, Libra, Scorpio, Sagittarius, Capricorn, Aquarius and Pisces) but the astronomical constellations are referred to here using their English equivalents (Ram, Bull, Twins, Crab, Lion, Virgin, Scales, Scorpion, Archer, Goat, Waterman and Fishes). The so-called 'thirteenth constellation', Ophiuchus or 'Serpent-Bearer', which lies between Sagittarius and Scorpio, was not included by Steiner.

Researching celestial formative forces

After Steiner's death in March 1925, experiments were conducted by his followers, including at the Goetheanum in Switzerland from 1930–35, to try to establish links between the growth of crops and the 'formative' forces exerted on them by the celestial bodies. In 1939, Lili Kolisko (1889–1976), who had attended Steiner's 1924 *Agriculture* course at Koberwitz, published *Agriculture of Tomorrow* with her husband Eugen. As well as looking at the relationships between the sun, moon and the visible planets (Mercury, Venus, Mars, Jupiter and Saturn) and certain metals, the Koliskos examined the moon's 29.5 day synodic, waxing–waning or new moon to full moon rhythm, and its effect on seed germination. Seeds tend to absorb the greatest amount of water on the days prior to the full moon, and Kolisko's conclusion was that seeds sown at or just before full moon showed the best germination rates and produced the highest yields. This appeared to back up Steiner's insistence that moon forces work strongly through the 'watery' element, and as seeds are watered into the ground, it would be natural for this potential lunar growth force to be communicated by the water.

Following on from Kolisko, a Swiss called Franz Rulni produced a biodynamic calendar in 1948. As well as recognizing the importance of the synodic full moon to new moon rhythm, Rulni also described the tropical month, or how the moon's arc across the sky moves up (ascending moon) and down (descending moon) in a 27.3-day cycle. Rulni had also attended Steiner's *Agriculture* course, and by 1952 was part of a biodynamic study group in Marburg, Germany in which the family of Maria Thun, who was to become the public face of lunar biodynamicism, also participated.

Thun initially used Rulni's calendar, noting that 'seeds germinated not only during the period of the waxing [synodic] moon, but for short [sidereal moon] periods grew much better, at other times less so'. Her research bore out the effect Rulni had noted, namely that crops transplanted during the descending moon phase required less watering than those transplanted during the ascending moon phase. Thun also picked up on Steiner's idea that plant growth was ruled by the four elements of earth, water, air and light/heat, and that these elements respectively influenced plant roots, leaves, flowers and fruit/seed. When biodynamic winegrowers talk of 'root days' and 'fruit days' they are referring to this aspect of Thun's research (see moon and stars: sidereal, below).

It should be stressed that biodynamic growers believe that the soil, which after all is the point where the cosmos meets the earth, must be alive and receptive to cosmic forces for the 'formative' forces Steiner talked about to be manifested in aspects of plant growth. Cosmic rhythms will be less obviously manifested in crops grown in soils upon which the biodynamic preparations 500–508 have not been used. This is why the prerequisite for Demeter biodynamic certification is regular use of the biodynamic preparations rather than working to celestial cycles.

Farmers who understand the rhythms and subtle influences of the wider universe on our planet might be considered more sensitive to the earth's own seasonal rhythms of autumn, winter, spring and summer. The idea of being able to plan agricultural work maybe months in advance according to celestial rhythms can be both appealing and reassuring, partly because it costs nothing, but mainly because it can allow farmers to set aside periods of an unfavourable moon in which to catch up on office chores.

LUNAR CYCLES

Moon and sun: synodic

The full moon to new moon cycle relates to the moon's position relative to the sun as seen from earth. It is the lunar cycle we are most familiar with because it governs which if any part of the moon is visible to us. This is due to the combined effect of the sun's light and the earth's shadow from different angles on the moon's surface.

Full moon to new moon

The full moon to new moon cycle lasts 29.5 days, during which the moon appears to increase in size when it is said to be waxing, and to decrease in size when it is said to be waning. The cycle begins with the new moon, when the moon and sun meet in the sky because the moon is right in front of the sun as viewed from earth. As the moon is reliant on sunlight to make it visible, the new moon is invisible from earth (except during a solar eclipse). Astronomers call new moon an example of a sun–moon conjunction, a conjunction occurring when two celestial bodies conjoin in the sky. The period marked by a planet returning to the same point in relation to the Sun is called synodic, after the Greek word *synodos* for meeting. The synodic period between new moons is the basis for our calendar month. The meeting of sun and moon at new moon symbolized copulation to the ancient Greeks, who saw new moon traditionally as a time of fertility and rebirth.

After several days, the waxing crescent of the moon becomes visible, with the left or western side of the moon illuminated in the southern hemisphere, and the right side in the northern hemisphere. Some interpret this crescent as resembling the upward pointing sprout of a newly germinated seedling. At first quarter, half the moon is visible, rising around midday and setting around midnight. The waxing moon is then said to become gibbous, with only a small portion in darkness. At full moon, the sun and moon stand opposite each other and the moon reaches its greatest extent as seen from the earth. As the moon rises, the sun sets. Astronomers call this an opposition, when celestial bodies stand 180 degrees opposite each other. Full moon is traditionally seen as a time of bounty, the small seedling having reached its full expression. Then the moon begins to wane and seems to diminish in size, initially waning gibbous and then moving into last (third) quarter when it rises at night and sets at midday. The cycle is complete when the waning crescent is invisible at new moon.

Biodynamic farmers believe the moon's influences on earth are felt through the medium of water. During the waxing moon there is an increase in the moisture content of the earth and in atmospheric humidity. This helps plant growth but may encourage fungal diseases. These can be kept in check by spraying horn silica 501 or common horsetail 508, and by avoiding opening the soil by ploughing (which may also encourage weed germination). On the plus side, soil moisture during the waxing moon period makes this a good time to spread solid compost, or apply dedicated soil sprays like horn manure 500 or Maria Thun's barrel compost 502–507, and even liquid manure sprays intended for the soil.

Farmer's lore holds that rain is likely at or soon after full moon. This may explain why years in which there are thirteen as opposed to the more usual twelve full moons are supposed to be lesser years for wine. The timing of the full moon may influence biodynamic farming decisions, encouraging growers to speed up grape picking, for example, in autumn, or to spray the vines pre-emptively to prevent fungal disease attack during the growing season.

Growers of annual above-ground crops from seeds are advised to sow a couple of days before full moon for better germination, based on the research by the Koliskos which followed Steiner's own recommendations for sowing and which has since been confirmed by Spiess (see moon and stars: sidereal, below). In contrast, a new moon is the appropriate time for vineyard cover crops to be ploughed into the soil so that they decompose, because this is when moon force activity is concentrated underground and exerts an inward contraction. For this reason new moon is also a beneficial time to cut hay (it dries more easily), sow seeds for root (below-ground) crops or do transplanting.

Moon and earth: apogee and perigee

This cycle relates to the varying distance of the moon from the earth. It takes 27.55 days for the moon to return to exactly the same place relative to earth as it is orbiting it. However, rather than orbiting earth in a perfect circle the moon does a boomerang-style ellipse. This means the moon is nearer earth during what is called perigee but farther away from it at what is called apogee, which is why this moon–earth cycle is also called the 'anomalistic' month.

Perigee

When the moon is at its closest point to earth at perigee, two things can happen. Tides are especially strong and there is more chance of a lunar landing. Neil Armstrong and his 1969 Apollo 11 mission team saved 42,000 kilometres of extra flying by going to the Moon at perigee.

The perigee moon is said to bring a 'winter' mood because the moon's 'watery' element strongly inhibits the sun's relationship with earth. Seeds sown at perigee may germinate poorly or grow to produce big yields, albeit of potentially 'watery' crops unsuited to long-term storage. This may suit growers of cut and come again salad, spinach and leaf beets more than it does winegrowers seeking age-worthy wines.

Biodynamic growers believe fungal disease spores causing mildew and rot become notably active at perigee. Spraying the common horsetail 508 preparation (see Chapter 2) a few days before perigee is a pre-emptive counter-measure, to force back fungal disease spores preparing to jump up from the soil and on to vine shoots, leaves and grapes. Spraying horn silica 501 (see Chapter 2) can also balance the force of the perigee moon by reinforcing the vine's relationship with the sun, especially if the grapes are ripening too slowly. Biodynamic winegrowers are especially attentive when perigee and full moon coincide (about once every fourteen or fifteen months). Ploughing during this 'watery' lunar double whammy might needlessly set soil-borne fungal-disease spores in motion while also releasing humidity from the newly turned soil.

Apogee

The moon's furthest point from earth or apogee is said to bring a 'summer' mood because tides are at their weakest, and the moon's 'watery' element is diminished. This makes it a good time to harvest fruit crops like wine grapes or sow potatoes. Root and leaf crops like carrots and lettuce sown at apogee tend to shoot up and run to seed (bolt) too quickly.

The 'watery' influence that full moon is said to bring is thought to be at its least potent if it coincides with apogee, although Soper cites Agnes Fyfe's experiments on the sap of mistletoe, hellebore and iris which suggested that the power of the full moon is actually weaker at perigee than at apogee, with exactly the reverse true for the new moon.[2]

As seen from earth, the 'speed' of the moon's elliptical orbit in front of the twelve star constellations on the ecliptic plane varies. This is why biodynamic sowing calendars show how the moon spends less time in front of a given star constellation at perigee than it does in front of the same constellation at apogee.

Moon and stars: sidereal

There are eighty-eight constellations visible in our sky, but the sun, moon and planets only pass in front of those lying along the ecliptic, as we have seen. Whereas the sun takes one year to pass in front of all twelve of them, it takes the moon only 27.3 days; and whereas the twelve signs in the tropical zodiac of popular astrology appear similarly sized (30 degrees) when laid out in a circle, the astronomical constellations visible in the sky vary in size, or width along the ecliptic, from Virgin (46 degrees) to Scales (21 degrees). The astronomical constellations form the sidereal zodiac, from the Latin word *sidera*, meaning star. Some constellations actually overlap, such as Goat and Waterman, and there are big gaps between others, such as Goat and Archer.

The point at which constellation boundaries are said to begin and end is based on the work of Ptolemy of Alexandria (c. AD100–78), a Greek who believed the earth was the centre of the universe, with the sun, moon and stars revolving around it. Ptolemy drew up a star map with 1,028 stars, and this was used by the International Astronomical Union (www.iau.org) when it demarcated the constellations in 1928. Slightly different astronomical divisions than the IAU are used by the biodynamic movement,[3] notably by its ideological HQ the Gotheanum in Dornach, Switzerland which publishes a celestial calendar annually, and by Germany's Maria Thun whose own calendar made her the world's most well-known biodynamic farmer until her death in April 2012 (see celestial calendars, below).

Maria Thun's four-element theory: root days and fruit days

Maria Thun's day job was market gardening and farming. Her tips on improving crop yields and quality by working to celestial cycles in general and the moon's sidereal cycle in particular are based on her own research studies, none of which as far as I know have been peer reviewed in scientific journals. Thun began documenting her research in 1952 by sowing radishes on four different days during a ten-day period. The radishes grew quite differently in size and shape, even though soil type, manuring and watering had all been identical (although Thun appeared not to take account of how weather fluctuations influenced her results).

By 1957, Thun was convinced that the four elements of earth, water, air and fire – as represented by four groups of constellations related to one single element each – strongly influenced the growth not just of plants but of particular parts (organs) of plants: roots, shoots/leaves, flowers and fruits. The constellations in each three-strong group formed a trigon or trine, laying at roughly 90-degree angles to each other, although rather than use the Babylonian system, whereby each of the twelve constellations had equal space (i.e. 30 degrees each) along the ecliptic, Thun used the unequal divisions of the sidereal zodiac as did the Gotheanum, meaning the smallest constellation (Scales) was assigned just 21 degrees and the largest one (Virgin) 46 degrees of the zodiacal pie. Thun's aim was to show that the moon's physical position in front of the unevenly sized zodiacal constellations was significant to how plants grew.

Nick Kollerstrom explains that the four-element theory was first expressed in the fifth century BC in Sicily by Empedocles.[4] The latter believed nature used the four elements of fire, wind, sea and stone to paint with, just as artists use four colours. The ancient Greeks also applied the four-element theory in their four temperaments, much used in their medical practice, defining people's characters by their related element.[5] What Thun was attempting to show via her research was how the four elements, plant growth and the moon–zodiac cycle were related. She said her inspiration in making the link was a 1957 lecture by Günther Wachsmuth who had been Steiner's personal assistant and had attended the 1924 *Agriculture* course.[6] Wachsmuth had written about the 'four ethers' as Steiner had described them but, like Steiner, without ever linking them to the moon's passage across the zodiac (see anthroposophy, p.4).

The four ethers

Steiner's four ethers or etheric formative forces are the warmth ether which works through the element of warmth or fire; the light ether which works through the element of light or air; the chemical ether which works through the element of water, also known as the tone or sound ether; and the life ether which works through the earth. These etheric formative forces influence the shape or form of the plant, rather than its substance.

Higher, seed-producing plants first develop by putting roots into the earth. The root system then supports the creation of shoots and leaves full of water. With enough leaves plants can reproduce by forming flowers which need light to open and air to carry their scent to attract and guide pollinators like bees and other insects for fertilization. Finally, the fruit relates to fire, meaning the heat or warmth of the sun which flowers need to contract and become plant seeds (like grape pips) and then to ripen so that when they fall to ground the next cycle of life can begin.

Thun set out to prove that the moon's movement through the sidereal zodiac catalysed the growth of particular parts ('organs' in biodynamic parlance) of plants as the roots, leaves, flowers and fruit/seeds found expression in the four elements.

Thun said her results showed that root crops like parsnips (and also onions and potatoes which grow in the earth but are not considered true root crops) performed best when sown, watered and hoed with the sidereal moon standing in the earth constellations of Bull, Virgin and Goat. Flowers did best with the sidereal moon in air/light constellations Twins, Scales and Waterman. Leaf crops like cabbage and lettuce performed best with the sidereal moon in water constellations Crab, Scorpion and Fishes. Fruit/seed crops performed best with the sidereal moon in the warmth constellations of Ram, Lion and Archer, with fruit crops corresponding especially to Ram and Archer and seed crops (cereals, sunflowers) corresponding especially to Lion. Thun also found root crops that were sowed, hoed and picked during leaf periods produced plants with big, leafy tops but only small roots.

Using Thun's suggestions, vines – which relate to the fruit–seed/ warmth constellations – should be worked when the sidereal moon stands in Ram, Lion or Archer. As explained above, the sidereal moon's passage through any constellation varies in length, being longest at apogee and shortest during perigee. For example, the sidereal moon stood in Archer for 2.5 days at lunar apogee in March 2009, but for only 1.54 days in front of the same constellation at lunar perigee in July 2005. Clearly only the smallest biodynamic vineyards, or those with the most committed owners, can claim to be able to time all work to this rhythm, especially since Thun says nodes, eclipses and some other celestial aspects whose cycles are independent of the sidereal rhythm are unfavourable for working with crops.

Pragmatic winegrowers following the Thun model will time key structural tasks of long-term consequence such as grafting, replanting or retraining (switching from say spur to cane pruning) to the most favourable sidereal periods. Ploughing the most weed-afflicted plots to the Lion–Goat rhythm should normally be possible (see Chapter 5, weed ashing). Modern vineyard machinery such as quad bikes, which are far quicker than tractors (and which compact the soil far less), can allow biodynamic sprays such as horn manure 500 and horn silica 501 to be applied very quickly and thus at opportune periods during the moon's sidereal phase, even on the largest vineyards.

Conveniently for winegrowers, sidereal fruit–seed/warmth periods are followed by root/earth periods. These would seem almost as equally beneficial for vines, as wines which take on more root/earth influence because the vines were worked during root/earth could potentially manifest a stronger *terroir* character. The three root/earth constellations Goat, Bull and Virgin are also among the largest, occupying a combined 110 degrees of the zodiac, with Virgin (46 degrees) and Bull (36 degrees) respectively the biggest and third biggest of all the zodiacal constellations. The fruit–seed/warmth constellations are fourth (Lion, 35 degrees), fifth equal (Archer, 30 degrees, tied with Scorpion) and tenth (Ram, 24 degrees). When combined, the six fruit–seed/warmth and root/earth constellations comprise 199 degrees (55 per cent) of the zodiac's 360-degree total.

Maria Thun's research questioned

Maria Thun's annual biodynamic sowing and planting calendar was first published in 1963 and is now written by her son Matthias. Contemporary biodynamic winegrowers probably spend more time consulting the Thun calendar than Steiner's *Spiritual Foundations for the Renewal of Agriculture* or the *Agriculture* course. However, both the methodology and relevance of the aspect of Thun's work concerning sidereal cycles has been questioned.

Of those who have tried to replicate Thun's results, the most thorough attempt was made by Hartmut Spiess for his doctoral thesis.[7] This was initiated with Thun's help and advice. It took Spiess six years longer than the two predicted to finish it and his conclusion was that the moon did consistently influence plant growth, but not through the sidereal moon in the way Thun had claimed. Like Kolisko and Steiner, Spiess saw the full moon as producing crops with higher yields. Speiss also found that lunar perigee, lunar nodes and planetary occultations (see eclipses, nodes, occultations and transits, below) had no negative effects, all of which ran counter to Thun's claims. Others, like Kollerstrom,[8] who worked from data taken during radish sowings by Colin Bishop on thirty-nine consecutive days in Cardiff in 1978 questioned whether Thun's initial research merely attempted to discern the basic fourfold phenomenon of the etheric formative forces exerted by the constellations on plant organs rather than actually trying to work out at which point on the ecliptic the moon had to be for each of the four formative forces (earth, water, air, warmth) to become active. In contrast to Thun, Kollerstrom favours the equal (Babylonian) division of the constellations along the sidereal zodiac into twelve 30-degree pieces.

Moon: transmitter or reflector?

Finally, Steiner himself was clear about the moon's role. Steiner said that while we tend to think of the moon as reflecting only sunlight and nothing more, in fact 'along with the Moonbeams, the entire reflected cosmos comes toward the Earth. The Moon reflects everything that comes toward it. In a certain sense, the whole starry heavens are reflected by the Moon and stream toward the Earth ... [like] a very powerful cosmic organizing force that radiates down from the Moon into the plants.' One interpretation is that Steiner saw the moon as blocking forces from whatever celestial body or constellation lay behind it, especially as he had suggested as such in a lecture he gave just before the 1924 *Agriculture* course.[9] Perhaps this is why concrete proof of the model advocated by Thun, in which the moon acts as a transmitter of forces of whichever sidereal constellation lies behind it, remains so apparently elusive – for both biodynamic believers and their sceptics.

The ascending and descending moon

The ascending moon and descending moon cycle is one of the easier lunar cycles to work to, as it allows winegrowers two periods of almost two weeks each in which to time specific tasks.

This cycle is known as the tropical month by astronomers, but in biodynamics growers refer to the ascending or 'spring–summer' moon and the descending or 'autumn–winter' moon cycles.

Just as the sun's position in the sky relative to the equator is never fixed, moving from its highest point in the sky each year at summer solstice to its lowest point at winter solstice, so the height the moon reaches above the equator also varies. This is due to the 5.5-degree angle at which the moon rotates the earth.

Like the sun and planets in our solar system, the moon passes in front of the twelve constellations of the zodiac as it traverses the ecliptic. The difference is that whereas for the sun there is a six-month gap between its winter low and summer high points above the horizon, the gap for the moon's 'summer' high and 'winter' low points is 27.3 days, or just 13.65 days between the start of each ascending or descending moon period.

This means it can be an ascending 'summer' moon in winter and a descending 'winter' moon in summer, as well as vice versa and any variation in between. It also means the moon can be ascending or descending irrespective of whether it is also full moon or new moon because this (synodic) cycle lasts 29.5 days.

The ascending–descending moon rhythm is also independent of how far the moon is physically distant from the earth because this apogee-perigee cycle is roughly 27.55 days long.

Ascending and descending periods in both hemispheres

The ascending and descending moon is the only lunar rhythm which differs between the two hemispheres. A descending or winter moon in the northern hemisphere is an ascending or summer moon in the southern hemisphere, and vice versa – in the same way the winter sun in Australia is the summer sun in London. So, in the northern hemisphere when the moon reaches its lowest point in the zodiac in front of Archer, its rising point shifts northeast and the setting point moves northwest during ascending moon. After the moon reaches its highest point, in Twins, it becomes descending, its arc in the sky becoming lower each day, its rising point shifting towards southeast and its setting towards southwest.

In the southern hemisphere, during the moon's movement from Archer (high in the sky looking north) to Twins (low in the sky looking north), the rising and setting points move northwards.

The ascending or 'spring-summer' moon

During an ascending moon the moon's arc passes from Archer via Goat, Waterman, Fishes, Ram, and Bull to Twins in the northern hemisphere and from Twins to Archer in the southern hemisphere. The ascending moon rhythm is said to invoke a spring–summer mood. The earth is said to breathe out, and plant growth is concentrated above the soil level, as life forces stream upwards from the roots. The upper parts of the plant fill with sap, vitality and aroma. Ascending moon periods are good times to spray horn silica 501, to sow cover-crop seeds (stronger germination), and to harvest flowers for display (longer-lasting blooms), or even Christmas trees (the needles take longer to fall). Tying down the newly pruned vine canes in early spring seems to work well under an ascending moon as canes are less likely to snap. Ascending moon periods are the time to pick grapes (for longer-lived wines). Maria Thun recommended that fruit being picked for storage is best harvested during an ascending moon, for it will keep longer without spoiling, so this is of especial interest to winemakers of *recioto* styles, in which the grapes are dried after picking for days or even months and must not spoil before they are pressed into wine.

The descending or 'autumn-winter' moon

During a descending moon the moon's arc passes from Twins via Crab, Lion, Virgin, Scales, Scorpion and back to Archer in the northern hemisphere and from Archer to Twins in the southern hemisphere. When the moon is descending the earth breathes in, and growth forces are concentrated underground as sap flows downwards into the roots. The descending or autumn–winter moon period lends itself to working with the soil and plant roots, by applying solid compost, spraying horn manure 500 and other soil sprays, or to cultivate the soil. The descending period is the best time to plant young vines and to undertake pruning, for valuable plant sap will now be safely concentrated in the roots. Biodynamic winemakers will also try to take account of the periods of the ascending and descending moon when working in the cellar: for instance by racking wine during the descending phase when sediment will be at its most compact, and wine aromas are less likely to dissipate.

Being attentive to the ascending–descending moon cycles can help when grafting vine scions onto rootstocks ready for the planting of new vineyards. The cutting for the vine scion will be taken during the moon's ascending phase, so that the sap force is concentrated in the upper part of the vine, meaning its shoots, so there is abundant sap in the scion for it to graft more strongly. Ideally, the union of the vine scion and rootstock should also coincide with the moon's ascending phase, so that sap from the rootstock rises up into the vine scion to reinforce the graft union. The newly grafted rootstock/vine will be planted out in spring during a descending moon, when the sap is concentrated in the root zone and earth forces are strongest in the earth. Even stronger rooting is encouraged by spraying rootlets with horn manure 500 or Maria Thun's barrel compost 502–507 before planting.

Moon-Saturn opposition

Oppositions occur when celestial bodies stand opposite each other in the sky at 180 degrees, the most obvious opposition occurring between the sun and moon every 29.5 days at full moon. Sowing crops around full moon can result in high yields but the risk is the moon's watery influence renders crops, as Peter Proctor says, 'soft and leggy. Those germinated when the moon and Saturn are in opposition have more strength [disease-resistance] and structure, and are likely to develop into strong plants that can resist disease and pests.'[10]

Rudolf Steiner outlined in his 1924 *Agriculture* course how plant growth was strongest when the two opposite force poles of lime/calcium and silica were in balance, which is the case when the Moon is in opposition to Saturn.

Steiner said the moon and the inner planets Mercury and Venus supported or enhanced the calcium or growth processes which build and regenerate the physical organism. They influence fertility – reproduction, propagation, germination – and substance or matter. In contrast, the outer planets – Jupiter, Saturn and Mars – supported the silica processes influencing plant form, shape, structure and sensory capacity (see Chapter 2, horn manure 500 and the lime–silica polarity). Thus when the Moon and Saturn stand on opposite sides of the earth, which occurs every 27.5 days, the forces they respectively exert ray into the earth from each direction and a synergy occurs.

Working with plants in the two days leading up to moon opposition Saturn produces stronger and more resistant growth (calcium) and quality (silica). Horn manure 500 sprayed at this time provokes strong soil activity and plant rooting. Horn silica 501 sprayed at this time augments plant resistance and crop quality. Common horsetail 508 sprayed at this time shows improved effectiveness at balancing soil–plant relationships if I could pack only one celestial cycle into my desert island suitcase, Moon–Saturn opposition would be it.

Mercury, Saturn and Jupiter

The three zodiac constellations of Ram, Lion and Archer are not the only celestial bodies potentially giving heat impulses and fruiting forces to vines (see moon and stars: sidereal, above). They are supported by two planets, Mercury and Saturn. The first has a shorter year than ours, and the other a year that is much longer (29.5 years). Joly says that Mercury influences sap circulation, accumulation and growth, and helps prevent weakness and disease while Saturn influences taste, maturity and concentration.[11] Jupiter also has a role, Joly says, in helping juice formation. When warmth planets like Mercury or Saturn stand in fruit–seed/warmth constellations their combined effects are supposed to be enhanced. This is especially so when Mercury and Saturn are in opposition, with the Earth at the centre of the 180-degree straight line they form.

Saturn entered the fruit–seed/warmth constellation of Lion on 1 September 2006 and stayed there for three years, bringing potentially beneficial warmth influences to winegrowers. Planting vines when both Saturn and the sidereal moon were in this fruit–seed/warmth constellation would have been especially beneficial in the northern hemisphere, because Lion is a descending moon constellation there.[12] On 1 September 2009 Saturn left Lion and moved into Virgin and would stay there for five years. Joly says that 'for distant planets like Saturn, it is sometimes worth waiting fifteen years for the best moment to plant'.

Eclipses, nodes, occultations and transits

Although the planets orbit the sun in approximately the same plane as the ecliptic, the moon's orbit is tilted at an angle of 5 degrees, being sometimes above it and at other times below. The two points where the moon crosses the ecliptic are called nodes, and nodes are notable for being the only points where eclipses can occur. An eclipse occurs only if the moon is in line with the earth and sun at the time of conjunction (new moon) or opposition (full moon). Solar eclipses can occur when the new moon is at a node because the moon is directly in front of the Sun. Lunar eclipses can occur when the full moon is at a node because the Earth's shadow falls on the moon. When the moon is ascending through the path of the ecliptic this is an ascending node, and when it is descending this is a descending node. A descending node can occur when the moon is ascending in the sky, but descending in relation to the sun's path. This is because the nodal cycle is slightly different in length to the ascending–descending cycle, 27.2 days for the nodal compared to 27.3 for the ascending–descending.

Nodes were also known as the dragon's head (ascending or north node) and the dragon's tail (descending or south node) and the nodal cycle is also called the draconic month. Maria Thun maintained all nodes, and by implication eclipses, are unfavourable moments for agricultural work, affecting plant growth and inhibiting germination of seeds sown during or close to the node, for example. Whether agricultural work should be stopped just an hour or two either side of a node or an actual eclipse, which Thun recommends, or for longer, perhaps for the whole day, appears unknown. The nodes move round the ecliptic circle in an 18.6 year rhythm called the nutation cycle; this is the time it takes the moon's nodes to complete one clockwise revolution along the ecliptic plane, opposite to the movement of the perigee. It has yet to be proved whether or how this affects the earth's climate.

Other periods avoided by some biodynamicists include occultations which occur when sun or the moon pass in front of and thus occult (hide) a planet; and transits, when the interior planets Mercury and

Venus pass directly in front of the Sun (something the other planets cannot do).

CELESTIAL CALENDARS

The earliest celestial calendars which appear to show the twenty-nine-day synodic (full to new moon) lunar cycle form part of the famous cave paintings found at Lascaux in western France and date from the Upper Paleolithic (15,000 years ago). Around 5,000 years ago the Egyptians devised their calendars on predictions that the Nile would always flood when the 'Dog Star' (Sirius) would appear in the eastern sky just before sunrise, and thus it was time to harvest. Greek writers like Hesiod (c.700 BC) also noted the link between Sirius and harvest time.

The most famous and widely used biodynamically oriented celestial calendar in the modern era is Maria Thun's annual *Aussaattage* which first appeared in 1963 and is published annually in her native Germany (Verlag). Thun's son Matthias collaborated with his mother on this calendar until her death, aged 89, in 2012. It is now published under his name in English as *The Biodynamic Sowing and Planting Calendar* (Floris), in Italian as the *Calendario delle Semine* (Editrice Antroposofica Milano), and in French as the *Calendrier des Semis* (Mouvement de Culture Bio-Dynamique). Another French calendar, Calendrier Lunaire Diffusion's *Calendrier Lunaire* (calendrier-lunaire.net) has especially clear, comprehensive diagrams of celestial movements. The most widely used calendar in the US, Kimberton Hills' *Stella Natura* (www.stellanatura. com), also carries articles on a variety of holistic themes written for both novices and experts. In the southern hemisphere the *New Zealand Biodynamic Farming and Gardening Calendar* (Biodynamic Farming and Gardening Association in New Zealand; www.biodynamic.org.nz) has useful diagrams and is a very stimulating read. Another widely used calendar used in the southern hemisphere is Brian Keats' *Antipodean Astro Calendar* (www.astro-calendar.com). In the UK, Nick Kollerstrom produces an engaging, easy to use, entry-level guide called *Gardening and Planting by the Moon* (plantingbythemoon.co.uk).

8

CERTIFICATION

Biodynamics is the world's oldest officially defined form of organic agriculture. A trademark was first registered for it in 1928 and the name chosen was Demeter, from *Da Mater* or 'Earth Mother', the mythological Greek goddess of fertility and protector of the fruits of the earth.

Demeter is a private rather than government trademark. Its standards for both winegrowing and winemaking are at least equivalent to national (e.g. US) or supranational (e.g. European Union) standards for organics: examples would be lower allowable limits for copper-based sprays (see below), and stricter rules on animal welfare. Demeter does not allow routine debeaking of poultry or dehorning of cows, for instance.

DEMETER INTERNATIONAL

The 'Demeter' trademark is administered by Demeter International, a non-profit organization based in Darmstadt, Germany. In 2016 Demeter International represented 5,000 farms with nearly 150,000 hectares of land in more than 45 countries.

Country member organizations of Demeter International of interest for wine include Demeter-Bund Österreich (www.demeter.at) in Austria, Association Demeter France (www.bio-dynamie.org) in France, Demeter e.V. (www.demeter.de) in Germany, Demeter Associazione Italia (www.demeter.it) in Italy, the Bio Dynamic Farming and Gardening Association in NZ Inc (www.biodynamic.org.nz) in New Zealand, Demeter Verband Schweiz (www.demeter.ch) in Switzerland, the Biodynamic Association (www.biodynamic.org.uk) in the UK, and the Demeter Association, Inc in the United States (www.demeter-usa.org) in the US.

It should be noted that Australia is not part of Demeter International because there the Demeter trademark was registered by Alex Podolinsky of the Bio-Dynamic Research Institute (see Chapter 5, Alex Podolinsky's prepared horn manure 500 + 502–507 spray for why).

DEMETER INTERNATIONAL'S BIODYNAMIC STANDARDS

Demeter International's *Standard for Biodynamic Farming* was first published in 1992, and is under constant revision. It is the baseline its biodynamic farmers and thus its winegrowers adhere to. Demeter certification programmes in Demeter International member countries can set higher standards than Demeter International's baseline if they wish, however. For example, in America the Demeter Association, Inc in the United States grants only organic certification (via its Stellar® organic trademark) rather than its full Demeter Biodynamic® certification trademark to grapes grown beneath or close to high-voltage power.

Wine is considered a transformed crop, meaning raw material (the juice of wine grapes) is transformed (fermented) into wine. Both aspects of wine production – the growing of the grapes, and the transformation of their juice into wine – are covered by certification rules as will be explained below.

ORGANIC CERTIFICATION AS A BIODYNAMIC BASELINE

Farms or vineyards seeking Demeter International's biodynamic certification must first satisfy the criteria for certified organic production to be considered eligible. So whereas an organic vineyard may not necessarily also be biodynamic, a biodynamic vineyard must always be at least organic.

The global baseline for organics follows guidelines laid out by the International Federation of Organic Agriculture Movements or IFOAM (ifoam.org). This non-profit umbrella organization for the global organic movement was founded in 1972. IFOAM's guidelines stipulate farming which avoids man-made or chemically synthesized products like soluble fertilizers, herbicides, pesticides, insecticides, acaricides, nematicides and fungicides, and systemic products which leave residues in the crop by penetrating the cell walls of crop plants.

IFOAM's organic standards also prohibit genetically engineered or genetically modified organisms ('GMOs'). Although genetically engineered vines, rootstocks, fermentation yeasts and enzymes have been developed, their commercial use in wine has been limited, mainly due to fears of a consumer backlash.

Global trade in organic products became easier in 2012 after the European Union, Canada and the United States agreed to recognize each other's organic standards as being equivalent. In 2015 the US and Switzerland made a similar deal. The EU also recognizes some other non-EU wine producing countries like Australia, New Zealand, Argentina and Switzerland as having equivalent organic standards to its own.

Because Demeter International is a private body with its own higher-than-organic rulebook, farms or vineyards seeking biodynamic status will either be certified by their local Demeter office if this is also a government-accredited certification body (as is the case for the Biodynamic Association in the UK), or the local Demeter office will ask an accredited certification body to do the inspection and complete a report on its behalf. In France, for example, Demeter France asks Ecocert France to perform certification inspections on its behalf.

Certification bodies gain accreditation under a set of international norms called ISO65 (EN45011 in Europe) administered by what is known as a 'competent authority', usually the national ministry of agriculture. By being certified as at least organic by an accredited body, Demeter products gain access to international markets such as those mentioned above.

CERTIFICATION IN PRACTICE

Normally the certifier's inspector will make an appointment with the vineyard owner so that, for example, maps of vineyard plots, receipts for vineyard sprays and cleaning materials, inventories of farm machinery and buildings (winery, barrel and bottle cellars) are ready. Tests (of soil, vine leaves) may be taken by the inspector to check for residues of banned substances. Every eighteen months or so an unannounced visit may be made, meaning one day's notice is given to the grower in order to avoid a wasted journey. Inspectors can advise growers which sprays are accepted under the organic rules, but they cannot advise on any spray's use, purchase or effectiveness. Inspectors usually inspect the same farm for no more than two successive years to avoid favouritism or bias. The inspector's vineyard report is then passed to a certification officer and it is for the latter, not the former, to decide whether the vineyard qualifies as biodynamic/organic. Winegrowers claiming to be biodynamic or organic but who are 'too busy to do the certification paperwork' may use this cuddly-sounding excuse because they have something to hide – I speak from experience, having worked for one such. Record keeping to the standard organic and biodynamic certifiers require is part of good farm management and, as New Zealand's James Millton of The Millton Vineyard in Gisborne says, 'it's probably the best business plan any farmer can do'. Certification is increasingly seen by wine importers, retailers and sommeliers as a key part of the marketing mix. Furthermore, countries in which state monopolies control the import of alcohol are increasingly favouring wines which are certified organic or biodynamic over their conventional counterparts.

CONVERSION TO ORGANICS, BIODYNAMICS

Land being converted from conventional farming must undergo a conversion period before crops from it can be described as organic or biodynamic. This organic conversion period is two years for annual crops like carrots, peas or potatoes and three years for perennial crops like vines, olives or apples. To be officially biodynamic another two years is required or, if the vineyard is converting directly from conventional to biodynamics, these two years may be allowed to run concurrently with the three-year organic conversion period.

THE BIODYNAMIC® TRADEMARK

Demeter International says that only wines made from grapes originating from vineyards certified by accredited Demeter International member

Table 3: Examples of maximum permitted levels of total sulphur dioxide (mg/L) in bottled wines under selected organic/biodynamic regimes

Residual sugar g/L	Demeter	European Union	European Union	Demeter	SIVCBD	FNIVAB	Nature et Progrès	Switzerland	Demeter	Demeter	USDA NOP	USDA NOP	USDA NOP
	International			France	France	France	France		US	US	US	US	US
Labelling/ logo	Made from biodynamic grapes	Non-organic	Organic wine	Made from bio-dynamic grapes	Biodyvin	Organic wine	Organic wine	Bourgeon Suisse	Biodynamic® wine	Made with biodynamic® grapes wine	Made with organically grown grapes	Organic wine (naturally occuring sulfites)	Organic wine – sulfite free
Red < 5 g/L	110	160	100	70	80	100	70	120	Limited	100	100	<10	None detectable
Red > 5 g/L	140	210	180	70	105	150	130	170	Limited	100	100	<10	None detectable
Pink < 5 g/L	110	210	150	90	105	120	90	120	Limited	100	100	<10	None detectable
Pink > 5 g/L	180	260	230	130	130	210	130	170	Limited	100	100	<10	None detectable
White < 5 g/L	140	210	150	70	105	150	130	170	Limited	100	100	<10	None detectable
White > 5 g/L	180	260	230	130	130	210	130	170	Limited	100	100	<10	None detectable
Semi-sparkling > 15 g/L	n/a	150	120	60	92-117	150	130	170	Limited	100	100	<10	None detectable
Fully sparkling < 15 g/L	n/a	100	70	60	75-117	100	60	120	Limited	100	100	<10	None detectable
Sweet white non-botrytised	250	300	270	130	150-200	250	150	170	Limited	100	100	<10	None detectable
Sweet white botrytised	360	400	370	200	200	360	150	170	Limited	100	100	<10	None detectable
Fortified	n/a	200	170	80	100	100	80	170	Limited	100	100	<10	None detectable

Key: SIVCBD is the Syndicat International des Vignerons en Culture Bio-Dynamique (logo 'Biodyvin'). FNIVAB (Federation Nationale Interprofessionnelle des Vins de l'Agriculture Biologique) is France's organic winegrower lobby. Nature et Progrès is France's oldest organic farm union. Switzerland's national organic logo is Bourgeon Suisse. USDA NOP = United States Dept of Agriculture's National Organic Program.

countries and which conform to Demeter International's winemaking standard (of which more below) may carry the Demeter logo or mention the word biodynamic.

Note that in America, the word Biodynamic® (with a capital 'B') is a registered certification mark of the Demeter Association, Inc. in the United States whilst the term Biodynamics® with an 's' is a registered trademark (rather than certification mark) of the Biodynamic Farming and Gardening Association.

Global use of the terms biodynamic and biodynamics is not the sole preserve of Demeter International, however.

In Australia use of the word biodynamic [sic] is determined by the state via its National Australian Standard for Organic and Biodynamic Produce. This means any accredited certification body in Australia can certify a vineyard to biodynamic standards because these are incorporated into the national organic standards and cover both grape growing and winemaking. However, neither the Australian state nor Demeter International control use of the word 'Demeter' in Australia because this was registered there by Alex Podolinsky of the Bio-Dynamic Research Institute (see Chapter 5, as mentioned above). Note, too, that Podolinsky uses the term Bio-Dynamic rather than biodynamic or Biodynamic.

SYNDICAT INTERNATIONAL DES VIGNERONS EN CULTURE BIO-DYNAMIQUE ('BIODYVIN')

As an alcoholic beverage, wine is seen by traditional biodynamic farmers as impeding one's spiritual or anthroposophical development (see Chapter 1, Steiner's path to anthroposophy, p.5). This caused a problem in the late 1980s and early 1990s in France as leading winegrowers there began practising biodynamics to improve wine quality.

Having joined the Association Demeter France to gain biodynamic certification, Marc Kreydenweiss of Domaine Kreydenweiss (Alsace) says, 'I did not feel at home. Some Demeter farmers who held very anthroposophical views seemed to feel that because as winemakers we made a[n alcoholic] product which ran counter to anthroposophy this made us demons. Others saw us winegrowers as being rich, much richer than them. They felt that if winegrowers started joining their biodynamic club then they would somehow lose out.'

As a consequence Kreydenweiss, Noël Pinguet of Domaine Huet (Vouvray), Véronique Cochran of Château Falfas (Côtes de Bourg) and others created the Syndicat International des Vignerons en Culture Bio-Dynamique (International Bio-Dynamic Winegrowers' Group) or SIVCBD (www.biodyvin.com). This has its own set of winegrowing rules and a certification logo, Biodyvin. The SIVCBD's rules broadly resemble Demeter International's in the vineyard but those of the European Union for winemaking to organic standards. Thus the SIVCBD allows the use of diammonium phosphate (DAP) whereas Demeter does not (see below).

Both Demeter France and the SIVCBD (since 2002) contract Ecocert France to conduct their certification audits. This makes it easier for those biodynamic vineyards who want to be members of both SIVCBD and Demeter France. Note that markets such as Sweden, for example, only recognize a product as 'biodynamic' if it is Demeter certified.

One key difference between Demeter International and the SIVCBD is that the SIVCBD awards its 'Biodyvin' certification trademark only to wines which have passed a peer-review approval tasting. Hence the SIVCBD judges a wine's 'biodynamicness' by its taste quality as well as by how it was grown and made. Demeter International does not impose an organoleptic peer review on its members' wines but simply says wines should come from grapes which are 'a true, unique expression of the vineyard individuality', and that the wine should express itself as 'preserved vitality'.[1] It is tempting to think that Demeter International is asking its winegrowers to make wines reflective of what the French call *terroir*, but in France *terroir* is essentially limited to soil type and site, altitude, type of slope, aspect to the sun, amount of annual rainfall and so on. For Demeter International, however, wine is also the product of what is going on much further above the winegrower's head than the clouds, meaning a wine made whilst 'working in cooperation with the rhythms of the cosmos', and encouraging 'the respectful combination of [both intangible] cosmic and material [e.g. annual rainfall or sunshine hours] forces'.

BIODYNAMIC WINE – GRAPE GROWING

Vineyards certified as biodynamic by Demeter (Australia excepted) conform to Demeter International's production standard. This stipulates that all the biodynamic preparations be used each year on the vines and other farmed areas. The preparations do not have to be applied to any permanently non-productive areas, like scrub or woodland surrounding the vineyard, for example.

Rudolf Steiner suggested biodynamic farms be as self-contained and thus as self-sufficient as possible.[2] This means that ideally all biodynamic preparations will be made on the farm, from plants and animal sheaths originating on the farm, from animals fed on food grown on the farm, and thus on land fertilized by the animals' own manure.

However, for winegrowers whose holdings often contain only perennial crops (the vines) the requirement to have their own animals is not obligatory as long as the land is treated with compost or teas based on animal manure, sown with green manures or cover crops to avoid bare soil, and the biodynamic preparations are used.[3]

Demeter accepts that it is not always possible for animals to be part of every holding (vineyards on very steep slopes, for example), but expects vineyard owners to show they have made an effort to incorporate animals in some way as part of their farming practice where possible (see composting raw materials, p.82). Even helping local apiculturists by hosting bee hives on the vineyard would stimulate new thought processes, leading winegrowers to create habitat breaks filled with flowering plants so the bees and other insects could feed, or to change how vine inter-rows were managed – perhaps by leaving the indigenous sward in every other row alone rather than ploughing it in.

The overriding principle is that vineyards should shed their inherent tendency to monoculture which perennial crops like vines encourage, to make better farms, better farmers and better surroundings for the community.

Winegrowers who do have their own livestock must compost the animal manure in compost piles into which the biodynamic compost preparations 502–507 are added. The compost must be matured correctly and spread on the land. If compost cannot be made then the six compost preparations 502–507 may be applied as sprays (see Chapter 3) which must be properly dynamized before use (see Chapter 4).

Other key aspects of Demeter International's standards[4] regarding the growing of biodynamic wine grapes concern various permitted agents, technologies and aids.

1. Permitted biological agents and technologies

Encouragement and use of natural control agents for plant pests (predator populations of mites, parasitic wasps, etc.). Beneficial insect predators can be encouraged by retention of native habitats around vineyards, or by selecting plants which host beneficial predators in cover-crop mixes, where sown. Seeds for cover crops must come from a certified biodynamic or organic source. If such seeds are unavailable then untreated seeds of conventional origin may be used (if approved). GMO seeds are not permitted.

Demeter allows beneficial typhlodrome mite predators to be introduced. Examples include *Neoseiulus californicus* which predates Chile's native red spider mite pest *Brevipalpus chilensis*, and *Metaseiulus occidentalis* which predates the Pacific spider mite in California. Demeter also allows parasitic wasps to be introduced, like *Anagrus* to control leafhoppers in California. As these introduced predators also need to be able to find alternate prey throughout their lifetime to reach the adult stage and reproduce, providing habitat for potential prey by careful cover cropping is an essential part of mite release.[5] As natural predators of spider mites may overwinter on two-year-old vine spurs (meaning spurs which produced the previous year's fruiting canes) cutting and then attaching these spurs strategically along supporting wires in plots known to be susceptible to mite attack helps restore a natural balance.

Insect traps. Coloured boards, sticky traps and attractants.

Pheromones. Sex attractants; attractants in traps and dispensers. These disrupt pests' mating cycles ('sexual confusion') preventing females of the target pest species from being mated, typically grape berry moths whose larvae pierce grape skins and so provide entry points for rot spores.

Mechanical repellents. Mechanical traps, slug and snail fences and similar methods.

Repellents. Non-synthetic agents to deter and expel pests, e.g. oil of thuya.

2. Permitted adhesion aids and materials to promote plant health

To promote plant disease resistance and inhibit pest and diseases, Demeter cites stinging nettle liquid manure, common horsetail 508 and wormwood teas as examples (see Chapter 6 for others), propolis, milk and milk products (used as anti-fungals), and waterglass (of which more below).

3. Permitted agents for use against fungal attack

The main diseases affecting both vine foliage and grapes are of fungal origin: powdery mildew/oidium (*Uncinula necator*), downy mildew (*Plasmopara viticola*) and bunch rot (*Botrytis cinerea*). As sap-penetrating systemic treatments are not permitted, contact barrier method treatments must be used. These include:

Wettable sulphur and dusting sulphur. Sulphur is the oldest recorded fungicide and was used in ancient Greece to control rust on wheat. It is used mainly to prevent powdery mildew. It is permitted in organics and biodynamics and is used by virtually all winegrowers worldwide. There are no restrictions on the amount used, although excess use may exacerbate soil acidification. Wettable rather than dusting sulphur may have to be used near centres of human population.

Waterglass. As sodium silicate or potassium silicate; used for powdery mildew.

Potassium bicarbonate. Used as a contact spray to break down the potassium-ion balance in the cells of the powdery mildew fungus causing them to shrink, then collapse. Good spray coverage is needed for it to be effective.

4. Permitted agents for pest control

Virus, fungal and bacterial preparations, e.g. Bacillus thuringiensis (BT) and granulose virus, which develop in and then secrete in the digestive system of grape worm caterpillars. These fatally affect the latter's development. Treatments are timed to coincide with the secondary breeding stage in the moth's population cycle, when the grape moth eggs are formed. BT is effective only when threat levels are low to medium. The threat is reduced by good canopy management, namely ensuring grape clusters are not overly shaded, especially early in the season. Climate data predicting attacks can also eliminate applications.

Pyrethrum. A contact insecticide based on pyrethrin extracted from a daisy, *Chrysanthemum cinariaefolium*. It is used against aphids and grape vine leafhopper vectors of Flavescence Dorée (grapevine yellows), phytoplasma disease. It must be sprayed in the evening as it is sensitive to ultraviolet light. It is also antagonistic to bees.

Quassia tea. An insecticide derived from the bark of the quassia (bitterwood) tree. It is used to control leaf rollers and aphids in fruit orchards.

Oil emulsions (without synthetic chemical insecticides) based on vegetable or mineral oil in the case of perennial crops. These are used to suppress powdery mildew, and to disrupt the feeding and reproduction cycles of mites and aphids.

Potassium soaps. Potassium salts ('soft soap') of fatty acids which disrupt the cuticle layer of aphids, mealy bugs, leafhoppers, whitefly, mites and other soft-bodied species.

Neem (azadirachtin, neem oil, neem oil soap). A non-synthetic botanical pesticide derived from the neem tree (*Azadiracta indica*), which grows in sub-tropical Asia. Neem is used to control mealy bugs, aphids, thrips and mites. It can also be used against powdery mildew in combination with powdered milk.

Rodenticide. Allowed only in bait boxes or similar, so that natural predators are not jeopardized.

5. Allowable aids on specialized crops, perennial crops (e.g. vines) and ornamental plants.

Copper-based anti-fungal sprays. Copper-based products in various formulations are allowed in organic farming because their use is deemed traditional and they are contact sprays, not systemics. They are used mainly for downy mildew, black rot and phomopsis (excoriose).

Despite being a trace element necessary for vine growth (at levels of 500 grams per hectare annually), copper can build up in the soil, causing toxicity. The risk of this is highest on lighter, sandier acid soils. Occurrences are rare and invariably indicative of poor management (e.g. calibrating sprayers wrongly, spraying in a high wind, failure to keep nozzles clean and unblocked, etc).

Copper is also damaging to soil organisms like earthworms, but its potentially harmful effects on them and micro-organisms (bacteria, yeast, fungi) can be offset by a combination of regular applications to the soil of compost, horn manure 500, allied to good sward and understorey management via cover cropping and sensible use of the plough.

Levels of copper can be reduced by combining it with teas from stinging nettle, common horsetail, willow, birch and other plants. These can make the copper more effective by reducing the hardening effect the copper has on vine leaves. Aerated compost teas (see p.131) can also be used as alternatives to copper.

Conventional vineyards can, of course, reduce their reliance on copper by also alternating it with systemic treatments at the risk of leaving residues in the wine.

Limits on copper for organic winegrowing in the European Union were 8 kilos per hectare per year until 2005, when they were reduced to 6 kilos per hectare per year which could be averaged over five years in perennial crops like vines. Limits are lowest of all for Demeter-certified biodynamic vineyards, at 3 kilos per hectare per year averaged over five years, ideally with a maximum of no more than 500 grams per hectare per spray. Germany has imposed this same 3 kilos per hectare per year limit on copper on all its winegrowers.

Growers in regions where downy mildew is present but who want to avoid copper altogether can plant interspecific vine crossings ('PIWIs') bred for their disease resistance. These varieties can be vigorous, requiring good (soil, canopy) management for an effective copper-free regime.

Copper can be applied in various formulations. They include Bordeaux mixture, which is copper sulphate mixed with calcium hydroxide (hydrated lime), copper hydroxide, copper oxychloride and cuprous oxide.

Sulphur preparations such as Hepar Sulphuris, and lime sulphur (calcium polysulphide) or *bouillie nantaise*. Lime sulphur can be sprayed in winter as a fungicide to clean (and harden) vine wood of fungal spores, especially phomopsis and powdery mildew. It is, however, antagonistic to overwintering spider and erinose mites (whether beneficial or otherwise). In Germany its use is no longer permitted for organics and thus for biodynamics.

Diatomaceous earth or DE is used as a contact insecticide. DE is a microscopically fine siliceous powder consisting of the fossilized shells of diatoms. Their abrasive edges cut insect pests, causing them to die of dehydration. DE is also used as a filtration medium in winemaking.

Plant material for new vineyards

The Demeter standards state that 'If plant material of the required varieties is available from Demeter production, then this must be used in preference. If plant material is available only from organic production, then this must be used.' In Europe, however, certified organic and biodynamic propagation material for new vineyards (rootstocks, budwood) and graft material in commercially registered nurseries must be treated with both conventional insecticides for *Schaphoïdus littoralis*, the leaf hopper vector of the phytoplasma disease Flavescence Dorée (grapevine yellows), and soil fumigants for nematode vectors of virus diseases.

In some European regions growers can get around this by saving both *V. vinifera* budwood and American rootstock material from their own vineyards and performing grafting operations themselves. Others have planted ungrafted vines, confident that their soil microbiology is benign, meaning complex and biodiverse yet phylloxera and nematode-free.

Commercially grown budwood may also have been created using meristem culture, a technique which biodynamic growers see as further disconnecting vines from their sexually reproductive origins.

WINE IS A TRANSFORMED PRODUCT

Regulators consider wine a transformed product. Wine exists if the raw material – the juice of freshly gathered grapes – is transformed into wine by alcoholic fermentation. Jam, butter, yoghurt, cider and beer are also considered transformed products.

The beginnings of the regulation of organic wine began in Europe after pioneering organic and biodynamic winegrowers there formed voluntary groups, initially as a means of exchanging information, then to draw up codes of practice. The most notable was 'Ecovin', the association of German organic growers formed in 1985. Its membership rules governed both vineyard (grape growing) and winery (winemaking, bottling).

European states later created a unified or single market via the European Economic Community (EEC), now renamed the European Union (EU). Its aim was to simplify cross-border trade between states by harmonizing regulations, and in 1991 a rule called (EEC) No. 2092/91 was drawn up for organic farming. However, this regulation covered specifically the grape-growing side of organic winegrowing, and not the winemaking side. The result was that wines from organic vineyards could only be labelled as 'wine from organically grown grapes' or 'wine from organic viticulture' rather than as 'organic wine', even if they had been transformed (fermented) with absolutely no fermentation aids (added yeast), additives (like sulphur dioxide preservative, of which more below) or agents (like finings to make wines taste smoother, or look brighter).

ORGANIC WINE LABELLING IN EUROPE

In 2012 the EU finally came up with a definition for 'organic wine' under Regulation (EU) No. 203/2012. This covered which inputs and practices were permitted for both grape growing and winemaking. The result was that any wine made from 100 per cent organic grapes made according to the EU's winemaking rule – which allowed the addition of sulphites – could be labelled 'organic wine'. The term 'wine made from organic grapes' was phased out. Producers of 'natural wine' (especially vociferous were European winegrowers in areas like the Loire, Beaujolais, the Veneto and Friuli) saw this as an opportunity to hammer home the ambiguity of the new EU legislation, under which consumers had no way of knowing whether an 'organic wine' had had sulphites added to it or not.

In the US, however, a clear divide between organic wines with added sulphites and those without had existed since 2002, when the US Department of Agriculture's National Organic Program (USDA NOP) came into force. This federal law on organics replaced the existing patchwork of organic legislation passed previously by individual states, such as California's 1990 Organic Food Act. It imposed a single organic baseline for all American farmers. It allowed wine producers – and consumers – to differentiate between 'wine from organic grapes' or 'organic wine', the differentiation revolving around sulphites, as will be explained below in the section on organic wine labelling in the US, p.174.

WHY ADD SULPHITES OR SULPHUR DIOXIDE TO WINE?

Sulphur dioxide (E220) is added to the vast majority of wine as well as to many foods as a disinfectant and preservative. In wine it prevents or slows micro-organisms like yeast and bacteria which (can, but not always) spoil the wine's taste – for example, by turning it to vinegar, wine being the mid-point between fresh grape juice and vinegar. Levels of sulphur dioxide or sulphites tend to be significantly lower in wine compared to foods like dried fruit or tinned soups, for instance; however, added sulphites are blamed for exacerbating the hangover effect, despite this being caused primarily by the effect on the body of excess alcohol combined with dehydration.

Nevertheless, the recent boom in interest in so-called 'natural' wines has seen increased demand for wines containing either low levels of added sulphites or none at all. Note, however, that the term 'natural wine' is an unofficial one. It is not permitted on wine labels in some EU countries, notably in Italy which is the EU's biggest wine producer, because 'natural' is not an independently regulated term unlike organics and biodynamics. It is also worth noting that wineries exhibiting at the plethora of 'natural wine' fairs and tastings that are now held in both hemispheres will contain both certified organic and biodynamic producers, as well as uncertified 'natural' ones, and that producers in all three camps will often offer both sulphited and unsulphited wine.

No added sulphites

An increasing if still small number of producers never add sulphites to any of their wines. Prerequisites of such successful minimal intervention in the winery are having the right grape variety planted on the right *terroir* (and where applicable, rootstock). Then often a high degree intervention in the vineyard is required, mainly of manual and ideally of animal labour: hand pruning, careful shoot positioning and bud-rubbing, and super-attentive but light-touch soil and sward management, ideally with horses or the like because their hooves and manure allow the soil to breathe and stay biodiverse. You must also learn when not to intervene and let nature take its course. This requires experience, observation, guts and raw talent, both in the vineyard and winery. And, as wine is a food product easily spoiled by microbes, good cellar hygiene is critical – good cellar hygiene meaning the winery should be clean rather than sterile so that ambient yeasts and other micro-organisms can work effectively. The most important ingredient of all, though, is healthy, ripe, clean grapes containing balanced sugars and acids, and with the right balance of nutrients for healthy ferments. Ideally the winery should be set up to make optimum use of ambient warmth and cold, and of gravity when racking from fermentation tank or barrel to ageing tank or barrel and thence to the bottling and storage areas.

Sulphite levels and wine labelling

The exact level of sulphites allowed in wines in general is dictated by the wine's colour, sweetness level and style (still, fizzy, fortified with grape spirit). Maximum sulphite levels for wine labelled as organic or biodynamic are set around 30 per cent lower than for conventional wines, on average. Allowable levels are lowest for red and dry wines, and highest for white wines and those with residual sugar.

Sulphite levels also have a bearing on exactly how wines are labelled, with some subtle but key differences between Europe and the US, and between organic and biodynamic.

ORGANIC WINE LABELLING IN THE US

In the US all organic, and by implication biodynamic, alcohol beverages carrying the USDA's organic seal must have met both the Alcohol and Tobacco Tax and Trade Bureau (TTB) and USDA organic regulations. These regulations differentiate between 'Wine made with Organically Grown Grapes' on the one hand and 'Organic Wine' on the other.[6]

Wines produced from certified organic grapes and which contain up to 100 milligrams per litre of sulphur dioxide added during winemaking are labelled 'Wine made with Organically Grown Grapes'. All grapes must be certified organic, but other ancillary agricultural aids (such as yeast or acid) are not required to be organic.

Wines produced from 100 per cent certified organic grapes, with no added sulphites and with less than 10 milligrams per litre of naturally occurring sulphites produced as a natural by-product of fermentation can be labelled 'Organic Wine' or even '100% Organic Wine'.

Wines produced from 100 per cent certified organic grapes, with no added sulphites and whose level of sulphites is so low as to be classed as 'none detectable' may also be labelled as 'Organic Wine – Sulfite Free'.

In addition, wines made from both non-organic and certified organic grapes can show the presence of the latter in the voluntary ingredients listing on the back label. For example: 'Ingredients: Organic Chardonnay grapes, Sauvignon Blanc grapes, Colombard grapes, sulfur dioxide. 75% organic ingredients.' Neither the front nor back label of such wines may carry the UDSA organic seal, however, and the term 'organic' is not permitted on the front label.

BIODYNAMIC WINE LABELLING

Biodynamic farming evolved out of anthroposophy or spiritual science (see Chapter 1), which proscribes alcohol because of its potential to diminish one's spiritual development. Until relatively recently so few Demeter farms worldwide were wine-grape vineyards there was little need for regulation to cover winemaking. This changed from the late 1980s onwards as increasing numbers of winegrowers, especially in France, began adopting biodynamics. They wanted clear guidelines on transforming raw biodynamic grapes into finished wine, but an absence of such guidelines saw the creation in France of the Syndicat International des Vignerons en Culture Bio-Dynamique or 'Biodyvin' (see above). It took until 2008 for Demeter International to produce its first global winemaking standard, and this is under constant revision.

BIODYNAMIC WINE LABELLING IN THE US

The Demeter Association, Inc in the United States offers two options when labelling wines.

Wines labelled 'Biodynamic® Wine' will come from 100 per cent Demeter-certified biodynamic grapes, and undergo minimal manipulation (including filtration) in the winery. They may contain small additions of added sulphites.

The second labelling category called 'Made with Biodynamic® Grapes Wine' denotes a wine that is made with 100 per cent Biodynamic grapes, but permits manipulations of the grapes as defined by the National Organic Program (NOP) 'Made with Organic grapes' category. This processing standard provides an option for 'Biodynamic® Wine' winemakers when the Biodynamic® Wine category cannot be achieved due to weather, etc.

DEMETER INTERNATIONAL'S BIODYNAMIC WINEMAKING STANDARDS[7]

Demeter International's Processing Standards now cover winemaking and begin by stating that Demeter biodynamic winegrowers should not see winegrowing purely as a means to an end, but more as a way 'to enrich the world and to celebrate the beauty of landscape and life'. The standards state that biodynamic methods should help each vineyard become an individuality in its own right. Wines should then be a true, unique, authentic expression of this individuality, what in wine-speak would be called '*terroir*-driven'.

However, Demeter is clear that its winemaking rules can only help minimize any degradation of the life forces which the grapes should have garnered from biodynamic vineyard management. The winemaker's touch should be as light as possible in terms of any energy, aids and additives used. Winemaking should simply be 'a rounding off of the processes underlying grape production in the vineyard'. Neither following the Demeter winemaking rulebook alone nor racking barrels by the moon will provide wines with such life forces. These only come from regular use in the vineyard of the biodynamic preparations 500–508.

Demeter asks that negative effects on the environment are minimized when making wine. Winery design in general is becoming greener and more carbon neutral against a backdrop of wild fluctuations in the price of fossil fuels and increased evidence of their role in contributing to climate change. Techniques for reducing energy use include gravity-fed and geothermal designs; green roofs and using rammed earth for better insulation; using wind and solar power; making more efficient use of water and waste water; using sustainable or renewable raw materials for construction. Technology is being developed, for example, in which algae capture fermentation gases like the main greenhouse gas, carbon dioxide, which are released during winemaking, and reusing or recycling them as biodiesel. Some winemakers offset their greenhouse gas emissions by planting a tree for every pallet of wine they ship or using power from local wind farms. The practice of offsetting has been criticized by environmentalists for failing to address the root cause of producing the greenhouse gases in the first place. Winemakers argue their grapes produce just as much carbon dioxide if left to decompose in the vineyard as they do during fermentation (assuming the wine ferments with no temperature control, electric pumping…).

Key aspects of Demeter International's wine processing standards for Biodynamic Wine are given below. Note that some Demeter International members, notably the Demeter Association, Inc in the United States and Association Demeter France, retained their own winemaking standards which pre-dated and can be slightly stricter than Demeter International's Processing Standards. France Vin Bio sets stricter rules than the EU minimum in some areas, and is France's leading organic winegrower group (formerly known as Fédération Nationale Interprofessionelle des Vins de l'Agriculture Biologique or FNIVAB).

E-numbers follow the International Numbering System for Additives.

The grapes

The grapes must be 100 per cent certified as biodynamic by Demeter.

Harvest

Demeter International prefers hand picking, but tolerates machine picking though this must be justified. EU and USDA NOP rules allow machine picking. Hand picking is obligatory for winegrower members of Nature & Progrès, France's oldest organic farmer association. 'Natural' winegrowers would pick by hand.

Harvest waste

Demeter International says pomace must be returned to the vineyard (via compost) if possible. (It may not make environmental sense to return pomace from biodynamic grapes sold and processed into wine a long way away from the vineyard.)

Winery equipment

Demeter International says maximum use should be made of gravity. Existing pumps that develop high shear or centrifugal forces can be used but are not permitted in new installations or when replacing machinery, and should be phased out.

Demeter International also says vats should be made from natural materials. Vats made from stainless steel, wood, stone, porcelain, clay (e.g. amphora) and concrete are allowed. See also 'oak ageing and barrels', below.

Demeter International says plastic tanks can be used temporarily when transferring wine, but not for storing wine. Nicolas Joly says wine cellars with too much metal (as stainless steel tanks) or even cement (whose metal content makes it a conductor) can, due to electrical radiation, produce electro-magnetic resonances and rhythms which impede the ability of 'living' wines to age to their full potential.[8] 'While the electric part of the electro-magnetic field can be easily resolved through a good earthing system … its magnetic part, which is more perverse, is sometimes impossible to cure', he says.

Cleaning and disinfection

Demeter International permits cleaning by water, steam, sulphur, soft soap, caustic soda, ozone, peracetic acid, acetic acid, hydrogen peroxide and citric acid, followed by flushing with water.

Temperature control and pasteurization

Demeter International allows tanks to be warmed to 35°C maximum during red wine fermentation. Use of heating and cooling to steer fermentation is permitted. Thermovinification and (flash) pasteurization are prohibited. France Vin Bio allows heating to 70°C.

Reverse osmosis and other techniques

Demeter International does not permit reverse osmosis (must or wine concentration), spinning cone (alcohol reduction), desulphiting of must (grape juice), ion exchange, electro-dialysis and cro-extraction. The EU permits but is trying to restrict or phase out ion exchange and desulphiting of must in organic grape concentrate.

Chaptalizing

Demeter International prefers no sugar addition to wines. It permits chaptalization up to a maximum 1.5% ABV. Sugar or grape juice concentrate should be from a Demeter source, but can be from an organic one. EU rules for organic wine state that organic sugar is preferred but not required. For sparkling wines, adding sugar (Demeter, if not organic) for tirage (secondary fermentation) is allowed up to a maximum of 1.5% potential alcohol.

Fortification

Demeter International stipulates biodynamic or organic grape spirit be used when fortifying. The EU now stipulates organic (or biodynamic) spirit be used when fortifying.

Yeast

Demeter International prefers wild (spontaneous) yeast ferments and allows a *pied-de-cuve* to be made from estate grapes. In the case of a justified stuck fermentation (<50 grams per litre unfermented sugar) Demeter International allows a cultured but neutral yeast to be added from a Demeter or organic source. If documented as unavailable, then yeast from a conventional (non-GMO) source is allowed (it must not have been grown on a petrochemical substrate or sulphite waste liquor). France Vin Bio allows only certified organic yeasts on which no coating of sorbitan monostearate (E491) has been applied.

Yeast nutrients

Yeast nutrients are added to juice or fermenting wine to prevent stuck fermentations. Demeter International allows yeast hulls and prefers them from a Demeter source, but permits organic ones. Demeter International says other yeast nutrients require prior approval. France Vin Bio allows yeast hulls, which must be organic if available.

Thiamin is allowed under EU organic rules. France Vin Bio deems thiamin unnecessary and so does not allow it.

Neither Demeter International nor the USDA NOP permit diammonium phosphate (DAP). DAP is allowed under EU organic rules (100 milligrams per litre) and by the Syndicat International des Vignerons en Culture Bio-Dynamique or SIVCBD but only in case of need, and on a case by case basis.

Enzymes (E440)

Demeter International does not permit pectolytic enzymes (for clarification) for wine made from grapes but does allow them from wines made from fruit like blackcurrants. The EU organic rule allows pectolytic enzymes as does France Vin Bio (<40 grams per hectolitre), with the latter preferring those from an organic source.

Ascorbic acid (E300)

Demeter International does not permit the use of ascorbic acid (Vitamin C), an antioxidant and sulphur dioxide enhancer. France Vin Bio follows the EU organic rules which permit ascorbic acid (25 grams per hectolitre) in organic grape must or wines.

Citric acid (E330)

Demeter International allows citric acid for cleaning purposes (see above), but not as an additive in wine (see acid adjustment via additions, below).

Sorbic acid (E200)

Use of sorbic acid, a yeast inhibitor, is not allowed by either Demeter International or the EU organic rule.

Sulphur dioxide

Demeter International says sulphur dioxide is to be restricted to the absolute minimum, but can be used. Table 3 (p.163) gives an indication of how permitted levels of sulphur dioxide vary for wines bearing ostensibly the same designation: an American 'organic wine' will contain no added sulphites but both a European 'organic wine' and an American 'Biodynamic® Wine' can and probably will.

Acid adjustment via temperature

Demeter International prefers tartrate stabilization, whereby unsightly but harmless tartrate crystals precipitate, to occur via natural cold stabilization – by, for example, leaving the cellar doors open during winter. However, Demeter International permits cold treatment (chilling).

Acid adjustment via additions

Demeter International prefers wines to be made without additions which provoke acid adjustment. However, it allows acid adjustment using tartaric acid (E334), calcium carbonate (E170) or $CaCO_3$, and potassium bicarbonate (E501(ii)) or $KHCO_3$ (all 150 grams per hectolitre max).

The EU organic rule and France Vin Bio allow additions of 100 grams per hectolitre of tartaric acid, citric acid, calcium carbonate, and potassium bicarbonate.

The EU also allows lactic acid and meta-tartaric acid (E353) but France Vin Bio does not allow either of these.

Potassium bitartrate (cream of tartar) is allowed both by the EU and France Vin Bio as a stabilizer.

Potassium alginate is allowed by the EU, and by France Vin Bio for clarification of sparkling wines only.

Calcium sulphate (gypsum) is allowed by the EU to acidify fortified wines.

Potassium tartrate (a de-acidifier) and cupric citrate (used to remove reductive flavours) are allowed by the EU, but not by France Vin Bio who deem them superfluous.

Oak ageing and barrels

Demeter International allows wooden barrels for oak ageing. Oak chips are allowed by the EU. The Demeter Association, Inc in the United States says oak character from barrels, staves or dust should not overpower the wine.

Malolactic fermentation

Demeter International prefers wines that go through malolactic fermentation to do so via indigenous bacteria, but allows lactic acid bacteria (GMO-free) addition. The EU and France Vin Bio allow lactic acid bacteria, the latter from an organic source if available.

Tannins

The EU organic rule allows additions of tannin, as does France Vin Bio, both specifying organic raw material if available. Not mentioned by Demeter; therefore not allowed.

Gum arabic (E414)

Gum arabic is used to stablize acid which might precipitate (drop out) of the wine after bottling and to improve mouthfeel. The EU organic rule allows additions of gum arabic (acacia gum) as powder or liquid, as does France Vin Bio, both specifying organic raw material if available. It is not mentioned by Demeter and therefore is not allowed.

Retsina wines

Demeter International says natural pine resin with no other aids or additives may be used in the production of traditional Greek Retsina wine.

Filtration

EU organic rules allow filtration above 0.2 microns only.

Demeter International allows filter pads made from cellulose, textile (chlorine-free) or polypropylene.

Demeter International permits diatomaceous earth (E551) as a filtration medium, as does the EU. The EU and France Vin Bio allow perlite, but this is not mentioned directly by Demeter International.

Fining

Demeter International allows egg-white fining with Demeter/organic egg whites only. It permits milk or milk products from a Demeter source and, if unavailable, casein (from an organic source). It also allows fining agents derived from pea, potato or wheat protein (organic if available).

France Vin Bio permits fining with fresh, frozen or powdered egg whites, or albumin if from organic sources.

Demeter International permits bentonite (tested for dioxin and arsenic), activated charcoal, and copper sulphate (<0.5 ppm).

France Vin bio permits bentonite, activated charcoal (but only for traditional method sparkling wines and at 50 grams per hectolitre, which is half the EU organic wine limit), and casein from an organic source if available.

France Vin Bio does not permit copper sulphate, whereas the EU allows it but its use is under revision.

Demeter International does not permit isinglass (E406), gelatine (E441) or blood.

France Vin Bio allows isinglass from an organic source where possible.

The EU organic wine rule allows gelatine derived from sheep or pigs, but due to fears over meat safety France Vin Bio does not permit it on the 'precautionary principle'.

Polyvinyl polypyrrolidone (E1202)

Neither Demeter International nor France Vin Bio permit polyvinyl polypyrrolidone (PVPP).

Potassium hexacyanoferrate

Demeter International, the EU and France Vin Bio do not allow potassium hexacyanoferrate (E536), also known as potassium ferrocyanide, to be used (it removes excess copper or iron from wine).

Racking

Demeter International prefers gravity to pumps when racking, but does permit micro-oxygenation (a form of static racking) but only to prevent reduction in the early phase of ageing. Winemakers following sidereal lunar rhythms (see Chapter 7) tend to rack and bottle when the descending moon stands in either a fruit–seed/warmth constellation (Ram in the southern hemisphere, Lion in the northern hemisphere, or Archer in either hemisphere) or a flower/light constellation (Waterman in the southern hemisphere, Scales in the northern hemisphere and Twins in either hemisphere). Others prefer to work only according to the weather and their barometers because the effect of how or if mineral or fruit smells are affected by timing racking and bottling according to specific celestial movements is not yet conclusive.[9]

Bottling aids

Demeter International permits sparging with carbon dioxide (E290) or CO_2 and nitrogen or N_2 (E941) at bottling. The EU also allows argon.

Bottles, packaging, seals, labels

Demeter International says glass bottles should be used (no stipulation as to maximum weight). Tins, PET and bag-in-a-box are not permitted (nor, by implication, are pouches or TetraPak).

Demeter International allows bottles to be sealed with natural cork, plastic corks, glass stoppers, screw caps and crown corks, or with a variety of tamper-proof seals (nirosta, plastic, tin, poly cap, sealing lacquer or wax). Demeter International does not stipulate whether the natural corks it favours should be sourced from FSC-accredited forests.

Demeter International's standards imply that wine labels should be from recycled or tree-free sources, and printed with chlorine-free processes using metal-free colour pigments; and that for shipping cardboard boxes with reduced cellulose content should be used. Biodegradable winery waste should be composted rather than being sent to landfill.

APPENDIX I – TASTING WINE BY THE MOON

Those who accept Maria Thun's contention that the sidereal moon (see Chapter 7) variously influences the vine's four 'organs' – its root system, its leaves, its flowers and its grapes – understandably feel this 27.3 day-long lunar rhythm also necessarily affects how wines taste too. Wines should thus show more obvious fruit/grape characters if tasted when the sidereal moon stands in the fruit–seed/warmth constellations Ram, Lion and Archer. Effects should be especially noticeable if the moon is also in its ascending phase – Ram in the northern hemisphere, Lion in the southern hemisphere and Archer in either – because the ascending moon should make wine aromas more pronounced as they 'ascend' or are pulled upwards out of the bottle or glass. During earth/root sidereal moon periods the same wine would be expected to show more obviously earthy characters, and during water/leaf periods wines might taste more vegetal or dilute, and so on.

Those keen to taste wine by the moon can consult Matthias Thun's *When Wine Tastes Best: A Biodynamic Calendar for Wine Drinkers* (Floris, UK), available in both book and app form.

When asked to organize wine tastings, I favour wines from vines with at least seven years' biodynamics on their CV so the full biodynamic effect has been felt strongly by both vines and winemaker. I also uncork the bottle or decant the wine twelve to forty-eight hours in advance. Biodynamic wines often need a little extra time to breathe or loosen up, coming from especially strong-rooted vines producing balanced grapes which are fermented with minimal intervention. Exposing the contents of the bottle to the elements allows any beneficial celestial influences to be fully expressed, too. Yet I find my appreciation of a wine is augmented more by good company and good food than by a good moon or favourable set of stars. And anyway, there are no great wines, only great bottles of wine, as the saying goes.

APPENDIX II – STUDYING BIODYNAMICS

Biodynamic training and education centres provide full- or part-time courses offering both formal biodynamic agricultural qualifications and informal training which prepare adult students to be able to run their own economically viable biodynamic farms. Would-be professional biodynamic farmers study practical aspects such as tractor driving, botany, geology, bookkeeping and the law, but the main focus is on making and using the biodynamic preparations, composting, understanding celestial cycles and the anthroposophical background to biodynamics. Rudolf Steiner founded his centre for spiritual scientific or anthroposophic research and learning at Dornach in Switzerland in 1913, naming it the Goetheanum (www.goetheanum.ch) after Johann Wolfgang von Goethe (see Chapter 1, Steiner's path to anthroposophy). The current building dates from 1928, three years after Steiner's death, the original having been destroyed in an arson attack.

Other biodynamic education centres include Emerson College (www.emerson.org.uk) in England, the Warmonderhof Training Centre (www.warmonderhof.nl) in the Netherlands and Taruna College (www.taruna.ac.nz) in New Zealand. This offers an Applied Organic and Biodynamic diploma which is part time (three eight-day semesters) for those in full-time jobs. The marked increase in interest in New Zealand for biodynamic winegrowing is largely due to the increased number of local viticulturists completing the course. New Zealand winegrower James Millton of The Millton Vineyard in Gisborne says his staff who take this diploma 'gain knowledge but as importantly undergo a positive change in character as individuals'. This bears out the notion that biodynamics is as important to the farmer as it is to the farm.

Most biodynamic centres of learning also offer courses for those aiming to teach Waldorf education which is based on Steiner's anthroposophical principles. Prospective Waldorf teachers must apprentice in biodynamic gardens. Hence, although America's leading Waldorf teacher education centre, Rudolf Steiner College (steinercollege.org) in Fair Oaks, California, offers no formal biodynamic agricultural training it does have its own biodynamic garden, the Raphael Garden. This opens in summer to members of the public who wish to study biodynamic gardening. Those wishing to learn how to make and use biodynamic preparations should join their local biodynamic farming and gardening association to be able to participate in weekend workshops.

NOTES

INTRODUCTION

1. Steiner, Rudolf, *Spiritual Foundations for the Renewal of Agriculture* (Bio-Dynamic Farming and Gardening Association USA, Inc. 1993), trans. by C. Creeger and M. Gardner, p.14

2. Courtney, Hugh, *What is Biodynamics?* (Steiner Books USA, 2005), p.10

3. A study undertaken by the University of Adelaide, Australia called '*The relative sustainability of organic, biodynamic and conventional viticulture*' was published in 2015. Its six-year trial in McLaren Vale compared conventional, organic and biodynamic viticulture, biodynamics meaning using only the two horn preparations 500 and 501, and excluding both the six compost preparations 502–507 and *Equisetum arvense* 508 or its Australia equivalent. The study concluded 'organic and biodynamic production led to improved soil quality, with more soil organisms and much greater earthworm populations. Wine quality was also improved, but in the absence of price premiums, this was achieved at a financial penalty to the grower through reduced yields and increased production costs.' Costs and benefits of the impact the different winegrowing systems had on the local environment were not taken into consideration.

4. See Giddens, Anthony, *The Politics of Climate Change* (Polity Press), 2006

5. Lorand, Andrew, 'Understanding Immunological Response Capacity', *Stella Natura Calendar 2000* (Kimberton Hills USA, 1999), p.26

6. See Costello, Michael J. and Daane, Kent M., 'Arthropods' (Chapter 8), in Ingels, C., Bugg, R., McGourty, G. and Peter Christensen, L.,

Cover cropping in vineyards (University of California publication 3338, 1998); and Roboz, Michael., 'Attracting Beneficial Insects' in *Biodynamic Perspectives* (New Zealand Biodynamic Association, 2001), ed. G. Henderson, pp.128–134

7.   Joly, Nicolas, *What is Biodynamic Wine?* (Clairview, 2007), trans. by M. Barton, p.77

CHAPTER 1

1.   Note the common horsetail 508 (*Equisetum arvense*) tea or liquid manure is not considered a true preparation by some biodynamic practitioners, even though they use it. This book does consider it such, and assumes therefore there are nine preparations in total.

2.   Stevens, Joseph A., *Applied Biodynamics* 33/2001, p.3

3.   Courtney, Hugh, 'Biodynamic Preparations', *Applied Biodynamics* 3/1993, pp.3–4

4.   Lovel, Hugh, *A Biodynamic Farm* (Acres USA, 2000), p.53

5.   Steiner, Rudolf, *Spiritual Foundations for the Renewal of Agriculture* (Rudolf Steiner Press, 1958), trans. by G. Adams, p.70

6.   Steiner, Rudolf, *Spiritual Foundations for the Renewal of Agriculture* (Bio-Dynamic Farming and Gardening Association, Inc. USA, 1993), trans. by C. Creeger and M. Gardner, p.35

CHAPTER 2

1.   The use of the Biodynamic preparations is permitted in Europe under article 12 (1) c) of EC regulation 834/2007.

2.   Demeter production standards for the *use of Demeter Biodynamic® and related trademarks June 2015* (Demeter-International e.V.), p.44

3.   Other names for horn manure include cow horn manure and horn humus, *la bouse de corne* or *le 500* (French), *das Hornmist-Präparat 500* (German), *cornoletame* (Italian), *preparación de boñiga en cuerno* or *preparado 500* (Spanish), and *chifre esterco* (Brazilian/Portuguese).

4.   Analyses reported in Baars, Ton and Pfirmann, Dorothee, 'On the effect of horn manure, discussion on evidence in accurate trials', *Star & Furrow* 115/Summer 2011, p.32 trans. by John Weedon show the pH of horn manure 500 to be neutral whether made in new horns (pH 7.0) or horns used once previously (pH 7.1).

5.	English wine producer Laverstoke Park Farm in Hampshire, which has its own soil testing laboratory, found its biodynamic horn manure 500 contained six times more bacteria, three times more fungi, eight times more protozoa and twice as many beneficial nematodes as its organic compost; and that in horn manure 500 the populations of active micro-organisms (bacteria, fungi, flagellates, amoebae, ciliates and nematodes) reached a clearly noticeable peak after being in the horn for precisely six months, and then subsequently declined. Dave Koball, who manages the extensive organic and biodynamic vineyards of Bonterra in California, says one way of thinking about horn manure 500 is 'as composting [cow manure] in a fermentation vessel [the cow horn] to produce a subset of micro-organisms that otherwise would not colonize your soil'.

6.	Steiner, Rudolf, *Spiritual Foundations for the Renewal of Agriculture* (Bio-Dynamic Farming and Gardening Association Inc. USA, 1993), trans. by C. Creeger and M. Gardner, pp.74–75

7.	Joly, Nicolas, *What is Biodynamic Wine?* (Clairview, 2007), trans. by M. Barton, p.61

8.	von Wistinghausen, C., Scheibe, W., von Wistinghausen, E., and König, U., *The biodynamic spray and compost preparations production methods* Booklet 1 (Biodynamic Agricultural Association UK, 2000), p.15

9.	See Begall, S., Červený, J., Neef, J., Vojtěch, O. and Burda, H., 'Magnetic alignment in grazing and resting cattle and deer', *Proceedings of the National Academy of Sciences of the United States* (August 2008), which suggests cows and deer align themselves with the earth's magnetic poles, showing these animals' acute sensory ability regarding forces. The authors 'challenge neuroscientists and biophysics to explain the proximate mechanisms'.

10.	Steiner, Rudolf, *Spiritual Foundations for the Renewal of Agriculture* (Bio-Dynamic Farming and Gardening Association, Inc. USA, 1993), trans. by C. Creeger and M. Gardner, p.71

11.	Steiner, Rudolf, *Spiritual Foundations for the Renewal of Agriculture* (Bio-Dynamic Farming and Gardening Association, Inc. USA, 1993), trans. by C. Creeger and M. Gardner, pp.74–5. See also von Wistinghausen, C., Scheibe, W., von Wistinghausen, E., and König, U., *The biodynamic spray and compost preparations production methods* Booklet 1 (Biodynamic Agricultural Association UK, 2000), p.15; and Poppen, Jeff, *The Barefoot Farmer* (USA, 2001), p.188

12. Bouchet, François, *L'Agriculture Bio-Dynamique* (Deux Versants, Paris, 2003), p.118

13. See Dewes, T., 'Das biologisch-dynamische Hornmistpräparat. Experimentelle Untersuchungen zur Beeinflussing seiner Wirkung auf Sommergersten-Keimpflanzen durch unterschiedliche Umweltbedingungen wahrend der Herstellungszeit.Teil 1: Mikrobiolo-gische Untersuchungen', *Lebendige Erde* 1983/1, pp.12–17.

14. Lovel, Hugh, *A Biodynamic Farm* (Acres USA, 2000), p.99

15. Baars *et al.*, ibid, suggest that horn manure 500 shows its biggest effect on soils 'where possibly relatively poor nutrient and life conditions prevail, or in winegrowing, where the vine grows on barren soil'.

16. Thornton Smith, Richard, 'Soil and light as a focus for biodynamics', *Star & Furrow* 123/2015, p.11

17. Steiner, Rudolf, *Spiritual Foundations for the Renewal of Agriculture* (Bio-Dynamic Farming and Gardening Association, Inc. USA, 1993), trans. by C. Creeger and M. Gardner, p.79

18. von Wistinghausen, C., Scheibe, W., von Wistinghausen, E. and König, U., *The biodynamic spray and compost preparations production methods* Booklet 1 (Biodynamic Agricultural Association UK, 2000), pp.17–18

19. Brinton, Will, 'Dynamic Chemical Processes underlying BD Horn Manure (500) Preparation', *Biodynamics* 214/1997

20. von Wistinghausen, C., Scheibe, W., von Wistinghausen, E. and König, U., *The biodynamic spray and compost preparations production methods* Booklet 1 (Biodynamic Agricultural Association UK, 2000), pp.16–17

21. A year before giving his 1924 *Agriculture* course Rudolf Steiner said feeding cows meat would turn them mad. No case of BSE or 'mad cow disease' has ever been found in a Demeter-certified biodynamic cow. Sceptics say this has nothing to do with the biodynamic cow's meatless diet because prion diseases like BSE are found in most mammals, are infectious and can jump species.

22. Steiner, Rudolf, *Spiritual Foundations for the Renewal of Agriculture* (Bio-Dynamic Farming and Gardening Association, Inc. USA, 1993), trans. by C. Creeger and M. Gardner, p.74

23. Masson, Pierre, *A Biodynamic Manual* (2nd edition 2014, Floris), pp.19–20

24. Courtney, Hugh, 'Further thoughts on making BD#500', *Applied Biodynamics* 9/1994, p.10

25.   Steiner, Rudolf, *Spiritual Foundations for the Renewal of Agriculture* (Bio-Dynamic Farming and Gardening Association, Inc. USA, 1993), trans. by C. Creeger and M. Gardner, p.79

26.   *Demeter production standards for the use of Demeter Biodynamic® and related trademarks June 2015* (Demeter-International e.V.), p.44

27.   Masson, Pierre, *A Biodynamic Manual* (2nd edition 2014, Floris), p.37

28.   Masson, Pierre, *A Biodynamic Manual* (2nd edition 2014, Floris), p.33

29.   Lovel, Hugh, *A Biodynamic Farm* (Acres USA, 2000), p.96

30.   Masson, Pierre, *A Biodynamic Manual* (2nd edition 2014, Floris), p.34

31.   Sattler, Friedrich, and von Wistinghausen, Eckard, *Bio-Dynamic Farming Practice* (Bio-Dynamic Agricultural Association UK, 1992), trans. by A. Meuss, p.93

32.   Poppen, Jeff, *The Barefoot Farmer* (USA, 2001), p.189

33.   Lorand, Andrew, 'Backyard Biodynamics', *Biodynamics* 254/2005, p.21

34.   Masson, Pierre, *A Biodynamic Manual* (2nd edition 2014, Floris), p.29

35.   Steiner, Rudolf, *Spiritual Foundations for the Renewal of Agriculture* (Bio-Dynamic Farming and Gardening Association, Inc. USA, 1993), trans. by C. Creeger and M. Gardner, p.88

36.   Other names for horn silica include *la silice de corne* or *la 501* (French), *das Hornkiesel-Präparat* (German), *il cornosilice* or *il preparato 501* (Italian), *preparación de sílice en cuerno* or *preparado 501* (Spanish), and *chifre silica* (Brazilian/Portuguese).

37.   Steiner, Rudolf, *Spiritual Foundations for the Renewal of Agriculture* (Bio-Dynamic Farming and Gardening Association, Inc. USA, 1993), trans. by C. Creeger and M. Gardner, p.74

38.   Thornton Smith, Richard, 'Soil and light as a focus for biodynamics', *Star & Furrow* 123/2015, p.12

39.   von Wistinghausen, C., Scheibe, W., von Wistinghausen, E. and König, U., *The biodynamic spray and compost preparations production methods* Booklet 1 (Biodynamic Agricultural Association UK, 2000), pp.24–25

40.   Joly, Nicolas, *What is Biodynamic Wine?* (Clairview, 2007), trans. by M. Barton, p.87

41.   Masson, Pierre, *A Biodynamic Manual* (2nd edition 2014, Floris), p.45

42.   Steiner, Rudolf, *Spiritual Foundations for the Renewal of Agriculture* (Bio-Dynamic Farming and Gardening Association, Inc. USA, 1993), trans. by C. Creeger and M. Gardner, p.82

43.   Steiner, Rudolf, *Spiritual Foundations for the Renewal of Agriculture* (Bio-Dynamic Farming and Gardening Association, Inc. USA, 1993), trans. by C. Creeger and M. Gardner, p.22

44.   Steiner, Rudolf, *Spiritual Foundations for the Renewal of Agriculture* (Bio-Dynamic Farming and Gardening Association, Inc. USA, 1993), trans. by C. Creeger and M. Gardner, p.74

45.   Bouchet, François, *L'Agriculture Bio-Dynamique* (Deux Versants, Paris, 2003), p.103

46.   Grahm, Randall, 'The Phenomenology of *Terroir*', *World of Fine Wine* 13/2006, p.107

47.   Grahm, Randall, 'The Phenomenology of *Terroir*', *World of Fine Wine* 13/2006, p.107

48.   Courtney, Hugh, 'Biodynamic Preparations', *Applied Biodynamics* 3/1993, p.3

49.   Steiner, Rudolf, *Spiritual Foundations for the Renewal of Agriculture* (Bio-Dynamic Farming and Gardening Association, Inc. USA, 1993), trans. by C. Creeger and M. Gardner, p.74

50.   Courtney, Hugh, 'BD #501 – The horn silica preparation', *Applied Biodynamics* 12/1998, p.4

51.   von Wistinghausen, C., Scheibe, W., von Wistinghausen, E. and König, U., *The biodynamic spray and compost preparations production methods* Booklet 1 (Biodynamic Agricultural Association UK, 2000), pp.26–7

52.   von Wistinghausen, C., Scheibe, W., von Wistinghausen, E. and König, U., *The biodynamic spray and compost preparations production methods* Booklet 1 (Biodynamic Agricultural Association UK, 2000), p.27

53.   von Wistinghausen, C., Scheibe, W., von Wistinghausen, E. and König, U., *The biodynamic spray and compost preparations production methods* Booklet 1 (Biodynamic Agricultural Association UK, 2000), p.27

54.   Bouchet, François, *L'Agriculture Bio-Dynamique* (Deux Versants, Paris, 2003), colour plates, p.iv

55.   Steiner, Rudolf, *Spiritual Foundations for the Renewal of Agriculture* (Bio-Dynamic Farming and Gardening Association, Inc. USA, 1993), trans. by C. Creeger and M. Gardner, p.74

56.   Moser, Patrick, 'JPI Summer Prep Seminar (June 26-28, 1998)', *Applied Biodynamics* 24/1998, p.8

57.   Masson, Pierre, *A Biodynamic Manual* (2nd edition 2014, Floris), p.45

58.   Courtney, Hugh, 'Brief directions for the use of biodynamic sprays 500, 501 and 508', *Applied Biodynamics* 50/2005, p.7

59.   Masson, Pierre, *A Biodynamic Manual* (2nd edition 2014, Floris), p.45

60.   *Demeter production standards for the use of Demeter Biodynamic® and related trademarks June 2015* (Demeter-International e.V.), p.44

61.   Masson, Pierre, *A Biodynamic Manual* (2nd edition 2014, Floris), p.50

62.   Masson, Pierre, *A Biodynamic Manual* (2nd edition 2014, Floris), p.52

63. Sattler, Friedrich and von Wistinghausen, Eckard, *Bio-Dynamic Farming Practice* (Bio-Dynamic Agricultural Association UK, 1992), trans. by A. Meuss, p.95

64. Courtney, Hugh, 'Summer in the biodynamic garden', *Applied Biodynamics* 8/1994, p.1

65. Courtney, Hugh, 'Brief directions for the use of biodynamic sprays 500, 501 and 508', *Applied Biodynamics* 50/2005, p.8

66. Masson, Pierre, *A Biodynamic Manual* (2nd edition 2014, Floris), p.49

67. Masson, Pierre, *A Biodynamic Manual* (2nd edition 2014, Floris), p.49

68. Masson, Pierre, *A Biodynamic Manual* (2nd edition 2014, Floris), p.48

69. Sattler, Friedrich, and von Wistinghausen, Eckard, *Bio-Dynamic Farming Practice* (Bio-Dynamic Agricultural Association UK, 1992), trans. by A. Meuss, p.94

70. Soper, John, *Bio-Dynamic Gardening* (Souvenir Press, 1996), eds. B. Saunders-Davies and K. Castelliz, pp.41–42

71. Sattler, Friedrich and von Wistinghausen, Eckard, *Bio-Dynamic Farming Practice* (Bio-Dynamic Agricultural Association UK, 1992), trans. by A. Meuss, p.95

72. Lorand, Andrew, 'Backyard Biodynamics', *Biodynamics* 254/2005, p.21

73. Lovel, Hugh, *A Biodynamic Farm* (Acres USA, 2000), p.100

74. Bouchet, François, *L'Agriculture Bio-Dynamique* (Deux Versants, Paris, 2003), p.101

75. Sattler, Friedrich and von Wistinghausen, Eckard, *Bio-Dynamic Farming Practice* (Bio-Dynamic Agricultural Association UK, 1992), trans. by A. Meuss, p.94

76. Courtney, Hugh, *What is Biodynamics?* (Steiner Books USA, 2005), p.24

77. Courtney, Hugh, 'Brief directions for the use of biodynamic sprays 500, 501 and 508', *Applied Biodynamics* 50/2005, p.8

78. Sattler, Friedrich and von Wistinghausen, Eckard, *Bio-Dynamic Farming Practice* (Bio-Dynamic Agricultural Association UK, 1992), trans. by A. Meuss, p.78

79. Masson, Pierre, *A Biodynamic Manual* (2nd edition 2014, Floris), p.116

80. von Wistinghausen, C., Scheibe, W., von Wistinghausen, E. and König, U., *The biodynamic spray and compost preparations production methods* Booklet 1 (Biodynamic Agricultural Association UK, 2000), p.74

81. Sattler, Friedrich and von Wistinghausen, Eckard, *Bio-Dynamic Farming Practice* (Bio-Dynamic Agricultural Association UK, 1992), trans. by A. Meuss, p.79

82. Lorand, Andrew, 'Backyard Biodynamics', *Biodynamics* 254/2005, p.21

83. Steiner, Rudolf, *Spiritual Foundations for the Renewal of Agriculture* (Bio-Dynamic Farming & Gardening Association, Inc. USA, 1993), trans. by C. Creeger and M. Gardner, p.128
84. Lorand, Andrew, 'Backyard Biodynamics', *Biodynamics* 254/2005, p.21
85. Masson, Pierre, *A Biodynamic Manual* (2nd edition 2014, Floris), p.119
86. Other names for (the) common or field horsetail (preparation) include *prêle de champs* (French), *Das Ackerschachtelhalm* or *Das Schachtelhalm-Präparat 508* (German), *equiseto* or *coda di cavallo* (Italian), *preparación de cola de caballo or preparado 508* (Spanish), and *cavalinho dos campos* or *cavalinha* (Brazilian/Portuguese).
87. Masson, Pierre, *A Biodynamic Manual* (2nd edition 2014, Floris), p.118
88. von Wistinghausen, C., Scheibe, W., von Wistinghausen, E. and König, U., *The biodynamic spray and compost preparations production methods* Booklet 1 (Biodynamic Agricultural Association UK, 2000), p.74
89. Lorand, Andrew, 'Backyard Biodynamics', *Biodynamics* 254/2005, p.21
90. Corrin, George, *Handbook on Composting and the Bio-Dynamic Preparations* (Biodynamic Agriculture Association UK, 2004), p.31
91. Lorand, Andrew, 'Backyard Biodynamics', *Biodynamics* 254/2005, p.21
92. Courtney, Hugh, 'Brief directions for the use of biodynamic sprays 500, 501 and 508', *Applied Biodynamics* 50/2005, p.9
93. Sattler, Friedrich and von Wistinghausen, Eckard, *Bio-Dynamic Farming Practice* (Bio-Dynamic Agricultural Association UK, 1992), trans. by A. Meuss, p.89
94. Masson, Pierre, *A Biodynamic Manual* (2nd edition 2014, Floris), p.118
95. Courtney, Hugh, 'Brief directions for the use of biodynamic sprays 500, 501 and 508', *Applied Biodynamics* 50/2005, p.8
96. Sattler, Friedrich and von Wistinghausen, Eckard, *Bio-Dynamic Farming Practice* (Bio-Dynamic Agricultural Association UK, 1992), trans. by A. Meuss, p.78
97. Steiner, Rudolf, *Spiritual Foundations for the Renewal of Agriculture* (Bio-Dynamic Farming and Gardening Association, Inc. USA, 1993), trans. by C. Creeger and M. Gardner, p.100
98. Steiner, Rudolf, *Spiritual Foundations for the Renewal of Agriculture* (Bio-Dynamic Farming and Gardening Association, Inc. USA, 1993), trans. by C. Creeger and M. Gardner, pp.91–93
99. See Soper, John, *Bio-Dynamic Gardening* (Souvenir Press, 1996), eds B. Saunders-Davies and K. Castelliz, p.42; and West, Lynette, 'Using Liquid Manures', *Star & Furrow* 109/2008, p.16
100. Steiner, Rudolf, *Spiritual Foundations for the Renewal of Agriculture* (Bio-Dynamic Farming and Gardening Association, Inc. USA, 1993), trans. by C. Creeger and M. Gardner, pp.101–102

101.   Proctor, Peter with Gillian Cole, *Grasp the Nettle* (Random House New Zealand, 1997), p.85

102.   Courtney, Hugh, 'Attention Readers', *Applied Biodynamics* 2/1992, p.7

103.   Jarman, Bernard, 'Christian von Wistinghausen 5th April 1933–20th August 2008', *Star & Furrow* 110/2009, p.27

104.   Podolinsky, Alex, *Bio Dynamic – Agriculture of the Future* (Bio-Dynamic Agricultural Association of Australia, 2000), p.17

105.   Steiner, Rudolf, *Spiritual Foundations for the Renewal of Agriculture* (Bio-Dynamic Farming and Gardening Association, Inc. USA, 1993), trans. by C. Creeger and M. Gardner, p.94

106.   Other names for (the) yarrow (preparation) include thousand seal, thousand leaf, noble yarrow, nosebleed, sanguinary and soldier's woundwort; *achillée* or *milles feuilles* or *achillée mille-feuille* (French), *Das Schafgarbe-Präparat 502* or *Schafgarbenblüten in Hirschblase präpariert* (German), *millefoglio* or *achillea* (Italian), *mil folhas* or *milefólio* (Portuguese), *milfolhas* or *milfolias* (Brazilian Portuguese) and *preparación de milenrama/mil en rama* or *preparado 502* (Spanish).

107.   von Wistinghausen, C., Scheibe, W., von Wistinghausen, E. and König, U., *The biodynamic spray and compost preparations production methods* Booklet 1 (Biodynamic Agricultural Association UK, 2000), p.30

108.   Proctor, Peter with Gillian Cole, *Grasp the Nettle* (Random House New Zealand, 1997), pp.70–71

109.   Lovel, Hugh, *A Biodynamic Farm* (Acres USA, 2000), p.107

110.   Steiner, Rudolf, *Spiritual Foundations for the Renewal of Agriculture* (Bio-Dynamic Farming and Gardening Association, Inc. USA, 1993) trans. by C. Creeger and M. Gardner, p.94

111.   Proctor, Peter with Gillian Cole, *Grasp the Nettle* (Random House New Zealand, 1997), p.69

112.   Lovel, Hugh, *A Biodynamic Farm* (Acres USA, 2000), p.107

113.   Lovel, Hugh, *A Biodynamic Farm* (Acres USA, 2000), p.107

114.   von Wistinghausen, C., Scheibe, W., von Wistinghausen, E. and König, U., *The biodynamic spray and compost preparations production methods* Booklet 1 (Biodynamic Agricultural Association UK, 2000), p.32

115.   Bouchet, François, *L'Agriculture Bio-Dynamique* (Deux Versants, Paris, 2003), colour plates, p.v

116.   In North America bladders from white-tailed deer or elk are sometimes used instead.

117.   von Wistinghausen, C., Scheibe, W., von Wistinghausen, E. and König, U., *The biodynamic spray and compost preparations production methods* Booklet 1 (Biodynamic Agricultural Association UK, 2000), p.31

118. von Wistinghausen, C., Scheibe, W., von Wistinghausen, E. and König, U., *The biodynamic spray and compost preparations production methods* Booklet 1 (Biodynamic Agricultural Association UK, 2000), p.31

119. Steiner, Rudolf, *Spiritual Foundations for the Renewal of Agriculture* (Bio-Dynamic Farming and Gardening Association, Inc. USA, 1993), trans. by C. Creeger and M. Gardner, p.96

120. Steiner, Rudolf, *Spiritual Foundations for the Renewal of Agriculture* (Bio-Dynamic Farming and Gardening Association, Inc. USA, 1993), trans. by C. Creeger and M. Gardner, p.96

121. Smith, Patricia, 'How to make the yarrow preparation', *Applied Biodynamics* 37/2002, p.4

122. Smith, Patricia, 'How to make the yarrow preparation', *Applied Biodynamics* 37/2002, p.6

123. von Wistinghausen, C., Scheibe, W., von Wistinghausen, E. and König, U., *The biodynamic spray and compost preparations production methods* Booklet 1 (Biodynamic Agricultural Association UK, 2000), p.33

124. Proctor, Peter with Gillian Cole, *Grasp the Nettle* (Random House New Zealand, 1997), pp.71–2

125. Proctor, Peter with Gillian Cole, *Grasp the Nettle* (Random House New Zealand, 1997), p.72

126. See Proctor, Peter with Gillian Cole, *Grasp the Nettle* (Random House New Zealand, 1997), p.72; and von Wistinghausen, Christian, Scheibe, Wolfgang, von Wistinghausen, Eckard, and König, Uli, *The biodynamic spray and compost preparations production methods* Booklet 1 (Biodynamic Agricultural Association UK, 2000), pp.33–4

127. von Wistinghausen, C., Scheibe, W., von Wistinghausen, E. and König, U., *The biodynamic spray and compost preparations production methods* Booklet 1 (Biodynamic Agricultural Association UK, 2000), p.35

128. Proctor, Peter with Gillian Cole, *Grasp the Nettle* (Random House New Zealand, 1997), p.72

129. von Wistinghausen, C., Scheibe, W., von Wistinghausen, E. and König, U., *The biodynamic spray and compost preparations production methods* Booklet 1 (Biodynamic Agricultural Association UK, 2000), p.35

130. Smith, Patricia, 'How to make the yarrow preparation', *Applied Biodynamics* 37/2002, p.9

131. Bouchet, François, *L'Agriculture Bio-Dynamique* (Deux Versants, Paris, 2003), colour plates, p.v–vi

132. Smith, Patricia, 'How to make the yarrow preparation', *Applied Biodynamics* 37/2002, p.9

133. Proctor, Peter with Gillian Cole, *Grasp the Nettle* (Random House New Zealand, 1997), p.72

134. Steiner, Rudolf, *Spiritual Foundations for the Renewal of Agriculture* (Bio-Dynamic Farming and Gardening Association, Inc. USA, 1993), trans. by C. Creeger and M. Gardner, p.96

135. *Demeter production standards for the use of Demeter Biodynamic® and related trademarks June 2015* (Demeter-International e.V.), p.44

136. Poppen, Jeff, *The Barefoot Farmer* (USA, 2001), p.196

137. Steiner, Rudolf, *Spiritual Foundations for the Renewal of Agriculture* (Bio-Dynamic Farming & Gardening Association, Inc. USA, 1993), trans. by C. Creeger and M. Gardner, pp.96–7

138. Proctor, Peter with Gillian Cole, *Grasp the Nettle* (Random House New Zealand, 1997), p.74

139. Other names for (the) chamomile (preparation) include *la camomille* (French), *Die Echte Kamille* or *Kamillenblüten im Rinderdünndarm präpariert* or *Präparat 503* (German), *camomilla* (Italian), *camomila* (Brazilian/Portuguese) and *preparación de manzanilla* or *preparado 503* (Spanish).

140. Proctor, Peter with Gillian Cole, *Grasp the Nettle* (Random House New Zealand, 1997), p.73

141. Proctor, Peter with Gillian Cole, *Grasp the Nettle* (Random House New Zealand, 1997), p.74

142. Lovel, Hugh, *A Biodynamic Farm* (Acres USA, 2000), p.108

143. *Demeter production standards for the use of Demeter Biodynamic® and related trademarks June 2015* (Demeter-International e.V.), p.45

144. See Proctor, Peter with Gillian Cole, *Grasp the Nettle* (Random House New Zealand, 1997), p.74; and von Wistinghausen, Christian, Scheibe, Wolfgang, von Wistinghausen, Eckard, and König, Uli, *The biodynamic spray and compost preparations production methods* Booklet 1 (Biodynamic Agricultural Association UK, 2000), pp.38–43

145. Bouchet, François, *L'Agriculture Bio-Dynamique* (Deux Versants, Paris, 2003), colour plates, p.vii

146. von Wistinghausen, C., Scheibe, W., von Wistinghausen, E. and König, U., *The biodynamic spray and compost preparations production methods* Booklet 1 (Biodynamic Agricultural Association UK, 2000), p.40

147. Proctor, Peter with Gillian Cole, *Grasp the Nettle* (Random House New Zealand, 1997), p.74

148. von Wistinghausen, C., Scheibe, W., von Wistinghausen, E. and König, U., *The biodynamic spray and compost preparations production methods* Booklet 1 (Biodynamic Agricultural Association UK, 2000), p.39

149.   von Wistinghausen, C., Scheibe, W., von Wistinghausen, E. and König, U., *The biodynamic spray and compost preparations production methods* Booklet 1 (Biodynamic Agricultural Association UK, 2000), p.39

150.   Proctor, Peter with Gillian Cole, *Grasp the Nettle* (Random House New Zealand, 1997), p.74

151.   Steiner, Rudolf, *Spiritual Foundations for the Renewal of Agriculture* (Bio-Dynamic Farming and Gardening Association, Inc. USA, 1993), trans. by C. Creeger and M. Gardner, p.225

152.   von Wistinghausen, C., Scheibe, W., von Wistinghausen, E. and König, U., *The biodynamic spray and compost preparations production methods* Booklet 1 (Biodynamic Agricultural Association UK, 2000), p.43

153.   Steiner, Rudolf, *Spiritual Foundations for the Renewal of Agriculture* (Bio-Dynamic Farming and Gardening Association, Inc. USA, 1993), trans. by C. Creeger and M. Gardner, pp.97–98

154.   von Wistinghausen, C., Scheibe, W., von Wistinghausen, E. and König, U., *The biodynamic spray and compost preparations production methods* Booklet 1 (Biodynamic Agricultural Association UK, 2000), p.43

155.   von Wistinghausen, C., Scheibe, W., von Wistinghausen, E. and König, U., *The biodynamic spray and compost preparations production methods* Booklet 1 (Biodynamic Agricultural Association UK, 2000), p.43

156.   Proctor, Peter with Gillian Cole, *Grasp the Nettle* (Random House New Zealand, 1997), p.75

157.   von Wistinghausen, C., Scheibe, W., von Wistinghausen, E. and König, U., *The biodynamic spray and compost preparations production methods* Booklet 1 (Biodynamic Agricultural Association UK, 2000), p.43

158.   von Wistinghausen, C., Scheibe, W., von Wistinghausen, E. and König, U., *The biodynamic spray and compost preparations production methods* Booklet 1 (Biodynamic Agricultural Association UK, 2000), p.39

159.   Bouchet, François, *L'Agriculture Bio-Dynamique* (Deux Versants, Paris, 2003), colour plates, p.viii

160.   *Demeter production standards for the use of Demeter Biodynamic® and related trademarks June 2015* (Demeter-International e.V.), p.44

161.   Other names for (the) stinging nettle (preparation) include *l'ortie piquante* (French), *Der Brennessel-Präparat 504* (German), *ortica piccante* (Italian), *urtiga maior* (Brazilian Portuguese), *urtigão* (Portuguese) and *preparación de ortiga* or *preparado 504* (Spanish).

162.   Joly, Nicolas, *What is Biodynamic Wine?* (Clairview, 2007), trans. by M. Barton, pp.13–14

163. Sattler, Friedrich, and von Wistinghausen, Eckard, *Bio-Dynamic Farming Practice* (Bio-Dynamic Agricultural Association UK, 1992) trans. by A. Meuss, p.77

164. Steiner, Rudolf, *Spiritual Foundations for the Renewal of Agriculture* (Bio-Dynamic Farming and Gardening Association, Inc. USA, 1993), trans. by C. Creeger and M. Gardner, p.99

165. von Wistinghausen, C., Scheibe, W., von Wistinghausen, E. and König, U., *The biodynamic spray and compost preparations production methods* Booklet 1 (Biodynamic Agricultural Association UK, 2000), p.46

166. Lovel, Hugh, *A Biodynamic Farm* (Acres USA, 2000), p.109

167. Steiner, Rudolf, *Spiritual Foundations for the Renewal of Agriculture* (Bio-Dynamic Farming and Gardening Association, Inc. USA, 1993), trans. by C. Creeger and M. Gardner, p.98

168. Steiner, Rudolf, *Spiritual Foundations for the Renewal of Agriculture* (Bio-Dynamic Farming and Gardening Association, Inc. USA, 1993), trans. by C. Creeger and M. Gardner, p.105

169. It seems some stinging nettle plants may also be monoecious, having both distinct zones of male and female parts on the same plant.

170. Steiner, Rudolf, *Spiritual Foundations for the Renewal of Agriculture* (Bio-Dynamic Farming and Gardening Association, Inc. USA, 1993), trans. by C. Creeger and M. Gardner, p.133

171. Bouchet, François, *L'Agriculture Bio-Dynamique* (Deux Versants, Paris, 2003), colour plates, p.ix

172. Steiner, Rudolf, *Spiritual Foundations for the Renewal of Agriculture* (Bio-Dynamic Farming and Gardening Association, Inc. USA, 1993), trans. by C. Creeger and M. Gardner, p.99

173. von Wistinghausen, C., Scheibe, W., von Wistinghausen, E. and König, U., *The biodynamic spray and compost preparations production methods* Booklet 1 (Biodynamic Agricultural Association UK, 2000), p.47

174. Steiner, Rudolf, *Spiritual Foundations for the Renewal of Agriculture* (Bio-Dynamic Farming & Gardening Association, Inc. USA, 1993), trans. by C. Creeger and M. Gardner, p.99

175. von Wistinghausen, C., Scheibe, W., von Wistinghausen, E. and König, U., *The biodynamic spray and compost preparations production methods* Booklet 1 (Biodynamic Agricultural Association UK, 2000), pp.47–8

176. von Wistinghausen, C., Scheibe, W., von Wistinghausen, E. and König, U., *The biodynamic spray and compost preparations production methods* Booklet 1 (Biodynamic Agricultural Association UK, 2000), pp.47–8

177. See Courtney, Hugh, 'The Michaelmas Preparation: BD#504 Stinging Nettle', *Applied Biodynamics* 24/1998, p.6

178. Steiner, Rudolf, *Spiritual Foundations for the Renewal of Agriculture* (Bio-Dynamic Farming and Gardening Association, Inc. USA, 1993), trans. by C. Creeger and M. Gardner, p.99

179. See Courtney, Hugh, 'New Insights on the Valerian Preparation – A Call to Examine Old Habits in Biodynamics', *Applied Biodynamics*, 33/2001, p.7

180. Steiner, Rudolf, *Spiritual Foundations for the Renewal of Agriculture* (Bio-Dynamic Farming and Gardening Association, Inc. USA, 1993), trans. by C. Creeger and M. Gardner, pp.99–100

181. Masson, Pierre, *A Biodynamic Manual* (2nd edition 2014, Floris), p.64

182. Moser, Patrick, 'JPI Summer Prep Seminar (June 26–28, 1998)', *Applied Biodynamics* 24/1998, p.8

183. Steiner, Rudolf, *Spiritual Foundations for the Renewal of Agriculture* (Bio-Dynamic Farming and Gardening Association, Inc. USA, 1993), trans. by C. Creeger and M. Gardner, p.45

184. Courtney, Hugh, *Applied Biodynamics* 24/1998, p.4

185. Courtney, Hugh, 'The Michaelmas Preparation: BD#504 Stinging Nettle', *Applied Biodynamics* 24/1998; and Courtney, Hugh, 'New Insights on the Valerian Preparation – A Call to Examine Old Habits in Biodynamics', *Applied Biodynamics*, 33/2001, p.7

186. Steiner, Rudolf, *Spiritual Foundations for the Renewal of Agriculture* (Bio-Dynamic Farming and Gardening Association, Inc. USA, 1993), trans. by C. Creeger and M. Gardner, pp.99–100

187. von Wistinghausen, C., Scheibe, W., von Wistinghausen, E. and König, U., *The biodynamic spray and compost preparations production methods* Booklet 1 (Biodynamic Agricultural Association UK, 2000), p.48

188. Proctor, Peter with Gillian Cole, *Grasp the Nettle* (Random House New Zealand, 1997), p.77

189. Courtney, Hugh, 'The Michaelmas Preparation: BD#504 Stinging Nettle', *Applied Biodynamics* 24/1998, p.6

190. von Wistinghausen, C., Scheibe, W., von Wistinghausen, E. and König, U., *The biodynamic spray and compost preparations production methods* Booklet 1 (Biodynamic Agricultural Association UK, 2000), p.48

191. Other names for (the) oak bark (preparation) include *l'écorce de chêne* or *la 505* (French), *der Eichenrinde-Präparat 505* or *Eichenrinde in Hirnhöhle eines Haustierschädels präpariert* (German), *corteccia di quercia* (Italian), *casca de carvalho* (Brazilian/Portuguese) and *preparación de corteza de roble* or *preparado 505* (Spanish).

192. Lovel, Hugh, *A Biodynamic Farm* (Acres USA, 2000), p.111

193. See König, Karl, *Earth & Man* (Bio-Dynamic Literature Wyoming USA, 1982), p.308

194. Lovel, Hugh, *A Biodynamic Farm* (Acres USA, 2000), p.111

195. Steiner, Rudolf, *Spiritual Foundations for the Renewal of Agriculture* (Bio-Dynamic Farming and Gardening Association, Inc. USA, 1993), trans. by C. Creeger and M. Gardner, p.101

196. Proctor, Peter with Gillian Cole, *Grasp the Nettle* (Random House New Zealand, 1997), p.77

197. Steiner, Rudolf, *Spiritual Foundations for the Renewal of Agriculture* (Bio-Dynamic Farming and Gardening Association, Inc. USA, 1993), trans. by C. Creeger and M. Gardner, p.101

198. Steiner, Rudolf, *Spiritual Foundations for the Renewal of Agriculture* (Bio-Dynamic Farming and Gardening Association, Inc. USA, 1993), trans. by C. Creeger and M. Gardner, p.101

199. Berger, Ed, 'Oak-Bark Preparation, Observing and learning from oak trees', *Star & Furrow* 102/2005, p.27

200. Proctor, Peter with Gillian Cole, *Grasp the Nettle* (Random House New Zealand, 1997), pp.78–79

201. Lovel, Hugh, *A Biodynamic Farm* (Acres USA, 2000), p.111

202. Poppen, Jeff, *The Barefoot Farmer* (USA, 2001), p.196

203. Courtney, Hugh, 'The oak bark preparation, organ of living thinking', *Applied Biodynamics* 42/2003 pp.10–11

204. von Wistinghausen, C., Scheibe, W., von Wistinghausen, E. and König, U., *The biodynamic spray and compost preparations production methods* Booklet 1 (Biodynamic Agricultural Association UK, 2000), p.51

205. von Wistinghausen, Christian, Scheibe, Wolfgang, von Wistinghausen, Eckard, and König, Uli, *The biodynamic spray and compost preparations production methods* Booklet 1 (Biodynamic Agricultural Association UK, 2000), p.51

206. Steiner, Rudolf, *Spiritual Foundations for the Renewal of Agriculture* (Bio-Dynamic Farming and Gardening Association, Inc. USA, 1993), trans. by C. Creeger and M. Gardner, p.111

207. Courtney, Hugh, 'How to make the oak bark preparation (BD#505)', *Applied Biodynamics* 42/2003, p.1

208. Masson, Pierre, *A Biodynamic Manual* (2nd edition 2014, Floris), p.129

209. Thun, Maria, *The Biodynamic Sowing and Planting Calendar 2005* (Floris, 2004), p.25

210. Proctor, Peter with Gillian Cole, *Grasp the Nettle* (Random House New Zealand, 1997), p.77

211. Joly, Nicolas, *Wine from Earth to Sky* (Acres USA, 1999) trans. by G. Andrews, p.38
212. Proctor, Peter with Gillian Cole, *Grasp the Nettle* (Random House New Zealand, 1997), p. 79
213. von Wistinghausen, C., Scheibe, W., von Wistinghausen, E. and König, U., *The biodynamic spray and compost preparations production methods* Booklet 1 (Biodynamic Agricultural Association UK, 2000), p.52
214. Masson, Pierre, *A Biodynamic Manual* (2nd edition 2014, Floris), p.113
215. Bouchet, François, *L'Agriculture Bio-Dynamique* (Deux Versants, Paris, 2003), colour plates, pp.xi–xii
216. The *Demeter production standards for the use of Demeter Biodynamic® and related trademarks June 2015* (Demeter-International e.V.), p.45 stipulates a skull from cows (<1 year old), pigs or horses.
217. von Wistinghausen, C., Scheibe, W., von Wistinghausen, E. and König, U., *The biodynamic spray and compost preparations production methods* Booklet 1 (Biodynamic Agricultural Association UK, 2000), p.53
218. Courtney, Hugh, 'The oak bark preparation, organ of living thinking', *Applied Biodynamics* 42/2003, p.8
219. Courtney, Hugh, 'How to make the oak bark preparation (BD #505)', *Applied Biodynamics* 42/2003 p.5
220. *Demeter production standards for the use of Demeter Biodynamic® and related trademarks June 2015* (Demeter-International e.V.), p.45
221. Steiner, Rudolf, *Spiritual Foundations for the Renewal of Agriculture* (Bio-Dynamic Farming and Gardening Association, Inc. USA, 1993), trans. by C. Creeger and M. Gardner, p.101
222. Courtney, Hugh, 'How to make the oak bark preparation (BD #505)', *Applied Biodynamics* 42/2003 pp.3–4
223. Steiner, Rudolf, *Spiritual Foundations for the Renewal of Agriculture* (Bio-Dynamic Farming and Gardening Association, Inc. USA, 1993), trans. by C. Creeger and M. Gardner, p.101
224. Steiner, Rudolf, *Spiritual Foundations for the Renewal of Agriculture* (Bio-Dynamic Farming and Gardening Association, Inc. USA, 1993), trans. by C. Creeger and M. Gardner, p.101
225. Proctor, Peter with Gillian Cole, *Grasp the Nettle* (Random House New Zealand, 1997), p.79
226. Bouchet, François, *L'Agriculture Bio-Dynamique* (Deux Versants, Paris, 2003), colour plates, pp.xi–xii
227. Courtney, Hugh, 'How to make the oak bark preparation (BD #505)', *Applied Biodynamics* 42/2003, p.6

228. Proctor, Peter with Gillian Cole, *Grasp the Nettle* (Random House New Zealand, 1997), p.80

229. *Demeter production standards for the use of Demeter Biodynamic® and related trademarks June 2015* (Demeter-International e.V.), p.44

230. Courtney, Hugh, 'Using the oak bark preparation', *Applied Biodynamics* 42/2003, p.7

231. Gardner, M.I., 'The innocent dandelion – a messenger from heaven', *Applied Biodynamics* 26/1999, p.10

232. von Wistinghausen, C., Scheibe, W., von Wistinghausen, E. and König, U., *The biodynamic spray and compost preparations production methods* Booklet 1 (Biodynamic Agricultural Association UK, 2000), p.61

233. Steiner, Rudolf, *Spiritual Foundations for the Renewal of Agriculture* (Bio-Dynamic Farming and Gardening Association, Inc. USA, 1993), trans. by C. Creeger and M. Gardner, p.103

234. Proctor, Peter with Gillian Cole, *Grasp the Nettle* (Random House New Zealand, 1997), p.80

235. Other names for (the) dandelion (preparation) include lion's tooth, priest's crown and swine's snout; *pissenlit* (French), *Löwenzahn-Präparat 506* or *Löwenzahnblüten im Rindergekröse präpariert* (German), *dente di leone* (Italian), *dente de leão* (Brazilian/Portuguese) and *preparación de diente de león* or *preparado 506* (Spanish).

236. von Wistinghausen, C., Scheibe, W., von Wistinghausen, E., and König, U., *The biodynamic spray and compost preparations production methods* Booklet 1 (Biodynamic Agricultural Association UK, 2000), p.63

237. Steiner, Rudolf, *Spiritual Foundations for the Renewal of Agriculture* (Bio-Dynamic Farming and Gardening Association, Inc. USA, 1993), trans. by C. Creeger and M. Gardner, p.103 & 135

238. Steiner, Rudolf, *Spiritual Foundations for the Renewal of Agriculture* (Bio-Dynamic Farming and Gardening Association, Inc. USA, 1993), trans. by C. Creeger and M. Gardner, p.272

239. von Wistinghausen, C., Scheibe, W., von Wistinghausen, E. and König, U., *The biodynamic spray and compost preparations production methods* Booklet 1 (Biodynamic Agricultural Association UK, 2000), pp.61–2 and 64

240. Steiner, Rudolf, *Spiritual Foundations for the Renewal of Agriculture* (Bio-Dynamic Farming and Gardening Association, Inc. USA, 1993), trans. by C. Creeger and M. Gardner, p.225

241. Proctor, Peter with Gillian Cole, *Grasp the Nettle* (Random House New Zealand, 1997), p.81

242. Bouchet, François, *L'Agriculture Bio-Dynamique* (Deux Versants, Paris, 2003), colour plates, pp.xiii–xiv

243. von Wistinghausen, C., Scheibe, W., von Wistinghausen, E. and König, U., *The biodynamic spray and compost preparations production methods* Booklet 1 (Biodynamic Agricultural Association UK, 2000), p.65

244. Bouchet, François, *L'Agriculture Bio-Dynamique* (Deux Versants, Paris, 2003), colour plates, p.xiv

245. Maria and Matthias Thun's *Biodynamic Sowing and Planting Calendar 2010* (Floris, 2009) describes Ram as being Mercury's 'favourite Warmth [i.e. fruit–seed/warmth] constellation', p.23

246. Proctor, Peter with Gillian Cole, *Grasp the Nettle* (Random House New Zealand, 1997), p.81

247. Courtney, Hugh, 'Dandelion preparation: mesentery or peritoneum?', *Applied Biodynamics* 16/1996, p.3

248. von Wistinghausen, C., Scheibe, W., von Wistinghausen, E. and König, U., *The biodynamic spray and compost preparations production methods* Booklet 1 (Biodynamic Agricultural Association UK, 2000), p.65

249. Proctor, Peter with Gillian Cole, *Grasp the Nettle* (Random House New Zealand, 1997), p.81

250. Lovel, Hugh, *A Biodynamic Farm* (Acres USA, 2000), p.112

251. *Demeter production standards for the use of Demeter Biodynamic® and related trademarks June 2015* (Demeter-International e.V.), p.44

252. Smith, Patricia, 'Using Valerian the way Steiner intended – An update', *Applied Biodynamics* 43/2003-4, p.7

253. Other names for (the) valerian (preparation) include *la valériane* (French), *Baldrian-Präparat 507* or *Baldrianblüten* (German), *valeriana* (Italian), *valeriana* (Brazilian/Portuguese) and *preparación de valeriana* or *preparado 507* (Spanish)

254. Davis, Ben, 'Valerian, A transcript of a talk given by Ben Davis at the IBIG conference in Edinburgh', *Star & Furrow* 104/2005, p.24

255. De Liefde, Bert, 'Production of the valerian preparation 507', *Harvests* 50/2 (1997), p.9. See also Pfeiffer, Ehrenfried, *Formative Forces in Crystallisation*, and *Sensitive Crystallisation Processes* (both published by Rudolf Steiner Press, London, 1936)

256. Bouchet, François, *L'Agriculture Bio-Dynamique* (Deux Versants, Paris, 2003), colour plates XV–XVI

257. Proctor, Peter with Gillian Cole, *Grasp the Nettle* (Random House New Zealand, 1997), p.84

258. Bouchet, François, *L'Agriculture Bio-Dynamique* (Deux Versants, Paris, 2003), colour plates XV–XVI

259. Courtney, Hugh, 'Valerian Experiment – A Call for Participants', *Applied Biodynamics* 31/2000–2001, p.3

260. Courtney, Hugh, 'Valerian Experiment – A Call for Participants', *Applied Biodynamics* 31/2000–2001, p.3

261. Davis, Ben, 'Valerian, A transcript of a talk given by Ben Davis at the IBIG conference in Edinburgh', *Star & Furrow* 104/2005, p.24

262. De Liefde, Bert, 'Production of the valerian preparation 507', *Harvests* 50/2 (1997), p.10

263. von Wistinghausen, C., Scheibe, W., von Wistinghausen, E. and König, U., *The biodynamic spray and compost preparations production methods* Booklet 1 (Biodynamic Agricultural Association UK, 2000), pp.69–70

264. Bouchet, François, *L'Agriculture Bio-Dynamique* (Deux Versants, Paris, 2003), colour plates, pp. xv–xvi

265. Proctor, Peter, with Gillian Cole, *Grasp the Nettle* (Random House New Zealand, 1997), p.84

266. Stevens, Joseph A., 'Prepared Valerian: The Secret of Finished Compost', *Applied Biodynamics* 33/2001, p.8

267. Proctor, Peter with Gillian Cole, *Grasp the Nettle* (Random House New Zealand, 1997), p.82

268. Stevens, Joseph A., 'Prepared Valerian: The Secret of Finished Compost', *Applied Biodynamics* 33/2001, p.9

269. Poppen, Jeff, 'Biodynamic farming', *Applied Biodynamics* 9/1994, p.9

CHAPTER 3

1. Joly, Nicolas, *What is Biodynamic Wine?* (Clairview, 2007) trans. by M. Barton, p.49

2. Steiner, Rudolf, *Spiritual Foundations for the Renewal of Agriculture* (Rudolf Steiner Press, 1958), trans. by G. Adams, p.70

3. Steiner, Rudolf, *Spiritual Foundations for the Renewal of Agriculture* (Bio-Dynamic Farming and Gardening Association, Inc. USA, 1993), trans. by C. Creeger and M. Gardner, p.27

4. Lovel, Hugh, *A Biodynamic Farm* (Acres USA, 2000), pp.86–87

5. Lovel, Hugh, *A Biodynamic Farm* (Acres USA, 2000), p.90

6. Steiner, Rudolf, *Spiritual Foundations for the Renewal of Agriculture* (Bio-Dynamic Farming and Gardening Association, Inc. USA, 1993), trans. by C. Creeger and M. Gardner, p.65

7. One reason why winter tree or pruning paste/wash is used. See Chapter 5.

8.   Corrin, George, *Handbook on Composting and the Bio-Dynamic Preparations* (Biodynamic Agriculture Association UK, 2004), pp.5–7

9.   Most studies show that cattle following the organic path of being fed on pasture and hay have far less risk of carrying *E coli* than cattle fed on concentrated food such as corn (maize), soy and ground-up animal products. See Sams, Craig, *The Little Food Book* (Alastair Sawday Publishing, 2003), p.47

10.   Masson, Pierre, *A Biodynamic Manual* (2nd edition 2014, Floris), p.59

11.   Joly, Nicolas, *What is Biodynamic Wine?* (Clairview, 2007) trans. by M. Barton, p.32

12.   Poppen, Jeff, *The Barefoot Farmer* (USA, 2001), p.147

13.   Koepf, Dr Herbert, *Compost – What It Is, How It Is Made, What It Does* (Biodynamic Farming and Gardening Association USA, 1980), p.5

14.   Masson, Pierre, *A Biodynamic Manual* (2nd edition 2014, Floris), pp.156–7

15.   Steiner, Rudolf, *Spiritual Foundations for the Renewal of Agriculture* (Bio-Dynamic Farming & Gardening Association, Inc. USA, 1993), trans. by C. Creeger and M. Gardner, p.91

16.   Proctor, Peter with Gillian Cole, *Grasp the Nettle* (Random House New Zealand, 1997), p.61

17.   Proctor, Peter with Gillian Cole, *Grasp the Nettle* (Random House New Zealand, 1997), p.60

18.   Corrin, George, *Handbook on Composting and the Bio-Dynamic Preparations* (Biodynamic Agriculture Association UK, 2004), p.9

19.   Proctor, Peter with Gillian Cole, *Grasp the Nettle* (Random House New Zealand, 1997), p.60

20.   See Diver, Steve, *Biodynamic Farming and Compost Preparation* (http://www.attra.ncat.org/attra-pub/PDF/biodynam.pdf)

21.   Corrin, George, *Handbook on Composting and the Bio-Dynamic Preparations* (Biodynamic Agriculture Association UK, 2004), p.11

22.   Thornton Smith, Richard, 'Soil and light as a focus for biodynamics', *Star & Furrow* 123/2015, p.10

23.   See König, Karl, *Earth & Man* (Bio-Dynamic Literature Wyoming USA, 1982)

24.   Lorand, Andrew, 'Backyard Biodynamics', *Biodynamics* 254/2005, p.21

25.   Courtney, Hugh, 'More on Biodynamic composting – II', *Applied Biodynamics* 11/1995, p.4

26.   See Thun, Maria, *Work on the Land and the Constellations* (Lanthorn Press, 1991), pp.28–29

27.   Lovel, Hugh, *A Biodynamic Farm* (Acres USA, 2000), p.114

28. *Demeter production standards for the use of Demeter Biodynamic® and related trademarks June 2015* (Demeter-International e.V.), p.44. See Table 2 in Chapter 2

29. Koepf, Dr Herbert, *Compost – What It Is, How It Is Made, What It Does* (Biodynamic Farming and Gardening Association USA, 1980), p.18

30. Bouchet, François, *L'Agriculture Bio-Dynamique* (Deux Versants, Paris, 2003), p.108

CHAPTER 4

1.   See Schiff, Michel, *The Memory of Water: Homoeopathy and the Battle of Ideas in the New Science* (Thorsons, 1995)

2.   See Schwenk, Theodor, *Sensitive Chaos* (Rudolf Steiner Press, 1996), trans. by O. Whicher and J. Weigley

3.   Pfeiffer, Ehrenfried, *Biodynamics: three introductory articles* (Biodynamic Farming and Gardening Association USA, 1956)

4.   Steiner, Rudolf, *Spiritual Foundations for the Renewal of Agriculture* (Bio-Dynamic Farming and Gardening Association, Inc. USA, 1993), trans. by C. Creeger and M. Gardner, p.73

5.   Masson, Pierre, *A Biodynamic Manual* (2nd edition 2014, Floris), p.36

6.   Dagostino, Kathryn, 'A new way of looking at water: an interview with Jennifer Greene', *Applied Biodynamics* 16/1996, p.6

7.   See Schwenk, Theodor, *Sensitive Chaos* (Rudolf Steiner Press, 1996), trans. by O. Whicher and J. Weigley

8.   Greene, Jennifer, 'The vortex in water and flowforms', *Stella Natura Calendar 2000* (Kimberton Hills USA, 1999), p.30

9.   Joly, Nicolas, *What is Biodynamic Wine?* (Clairview, 2007), trans. by M. Barton, p.62

10. Greene, Jennifer, 'The vortex in water and flowforms', *Stella Natura Calendar 2000* (Kimberton Hills USA, 1999), p.30

11. West, Lynette, 'Using Liquid Manures', *Star & Furrow* 109/2008, pp.15–16

12. Soper, John, *Bio-Dynamic Gardening* (Souvenir Press, 1996), eds. B. Saunders-Davies and K. Castelliz, p.40

13. Masson, Pierre, *A Biodynamic Manual* (2nd edition 2014, Floris), p.34

14. Steiner, Rudolf, *Spiritual Foundations for the Renewal of Agriculture* (Bio-Dynamic Farming and Gardening Association, Inc. USA, 1993), trans. by C. Creeger and M. Gardner, p.74

15. Steiner, Rudolf, *Spiritual Foundations for the Renewal of Agriculture* (Bio-Dynamic Farming and Gardening Association, Inc. USA, 1993), trans. by C. Creeger and M. Gardner, p.76—77

16. Baars, Ton, and Pfirmann, Dorothee, 'On the effect of horn manure, discussion on evidence in accurate trials', *Star & Furrow* 115 (Summer 2011), p.32 trans. by John Weedon

17. Storch, Stephen, 'Developing a hydraulic stirring machine', *Applied Biodynamics* 20/1997, p.8

18. See Beckett, Neil, 'Biodynamo', *Harpers Wine & Spirit Weekly*, 25 October 2002, p.31

19. Proctor, Peter, with Gillian Cole, *Grasp the Nettle* (Random House New Zealand, 1997), p.46

20. Storch, Stephen, 'Developing a hydraulic stirring machine', *Applied Biodynamics* 20/1997, p.8

21. Courtney, Hugh, 'Stirring vessels and sprayers (part 2)', *Applied Biodynamics* 20/1997, p.5

22. Steiner, Rudolf, *Spiritual Foundations for the Renewal of Agriculture* (Bio-Dynamic Farming and Gardening Association, Inc. USA, 1993), trans. by C. Creeger and M. Gardner, p.77

23. Wilkes, Thomas and Schwuchow, Jochen, 'John Wilkes 1930–2011', *Star & Furrow* 116/2012, pp.48–49

24. See Wilkes, John, *Flowforms – The Rhythmic Power of Water* (Floris, 2003)

25. Masson, Pierre, *A Biodynamic Manual* (2nd edition 2014, Floris), p.33

26. Masson, Pierre, *A Biodynamic Manual* (2nd edition 2014, Floris), p.112

27. Steiner, Rudolf, *Spiritual Foundations for the Renewal of Agriculture* (Bio-Dynamic Farming and Gardening Association, Inc. USA, 1993), trans. by C. Creeger and M. Gardner, p.83

CHAPTER 5

1. Schilthuis, Willy, *Biodynamic Agriculture* (Floris, 2003), trans. by T. Langham and P. Peters, p.42

2. See Marcel, Christian, 'Sensitive Crystallization, Visualizing the Qualities of Wines' (Floris, 2011), trans. by C. Moore

3. See comments by Chris Stearn in Courtney, Hugh and Stearn, Chris, 'Dr. Pfeiffer's BD compost starter and BD field spray', *Applied Biodynamics* 6/1993, p.10

4. Courtney, Hugh, *What is Biodynamics?* (Steiner Books USA, 2005), p.13

5. See comments by Courtney in Courtney, Hugh and Stearn, Chris, 'Dr. Pfeiffer's BD compost starter and BD field spray', *Applied Biodynamics* 6/1993, p.10; and Courtney, Hugh, 'Compost or Biodynamic compost', *Applied Biodynamics* 9/1994, p.12

6. Courtney, Hugh, 'Brief directions for the use of biodynamic sprays 500, 501 and 508', *Applied Biodynamics* 50/2005, p.9

7. Bouchet, François, *L'Agriculture Bio-Dynamique* (Deux Versants, Paris, 2003), p.67

8. Masson, Pierre, *A Biodynamic Manual* (2nd edition 2014, Floris), p.78

9. See Thun, Maria, *Work on the Land and the Constellations* (Lanthorn Press, 1979)

10. Masson, Pierre, *A Biodynamic Manual* (2nd edition 2014, Floris), p.79

11. Some practitioners chose to add the valerian 507 later and only when the barrel compost has transformed into humus. See the profile of valerian 507, Chapter 2

12. Henderson, Gita, 'Cow Pat Pit, Where Did it Come From?', *Harvests* 55/2 (2002), p.4

13. Wine from certified organic/biodynamic vineyards either produced or intended for sale in the US must now comply with the USDA's NOP (National Organic Program) restriction which prohibits the application of raw manure to land within ninety days prior to harvest for crops for human consumption not in direct contact with the soil or soil particles. Therefore, chickens must be removed from vineyards ninety days before harvest, although they are free to remain on other parts of the same property. The USDA's NOP regulations do not say how winegrowers should deal with any droppings left by birds overflying the vines in the ninety-day period before harvest takes place.

14. Henderson, Gita, 'Cow Pat Pit, Where Did it Come From?', *Harvests* 55/2 (2002), p.3

15. Lovel, Hugh, *A Biodynamic Farm* (Acres USA, 2000), p.123

16. Poppen, Jeff, *The Barefoot Farmer* (USA, 2001), p.193

17. Goldstein, W.A. and Barber, W., 'Yield and Root Growth in a Long-Term trial with Biodynamic Preparations'. *The proceedings of the International Society for Organic Farming Research (ISOFAR), Scientific Conference in cooperation with the International Federation of Organic Agriculture Movements (IFOAM) 21–23 September 2005, in Adelaide, South Australia.*

18. Bouchet, François, *L'Agriculture Bio-Dynamique* (Deux Versants, Paris, 2003), pp.158–9

19. Proctor, Peter with Gillian Cole, *Grasp the Nettle* (Random House New Zealand, 1997), p.113

20. Sattler, Friedrich and von Wistinghausen, Eckard, *Bio-Dynamic Farming Practice* (Bio-Dynamic Agricultural Association UK, 1992), trans. by A. Meuss, p.89

21. Bouchet, François, *L'Agriculture Bio-Dynamique* (Deux Versants, Paris, 2003), pp.120–121

22. Podolinsky's preferred name for his preparation is used here, but elsewhere prepared horn manure 500 + 502–507 is used for it to indicate exactly which seven biodynamic preparations it contains.

23. See Bouchet, François, *L'Agriculture Bio-Dynamique* (Deux Versants, Paris, 2003), p.133 and 162–3

24. Courtney, Hugh, 'Sequential spraying – Illusion, remarkable coincidence or reality?', *Applied Biodynamics* 6/1993, p.5

25. Steiner, Rudolf, *Spiritual Foundations for the Renewal of Agriculture* (USA, 1993), trans. by C. Creeger and M. Gardner, p.122

26. Wright, David, 'Possum Peppering and Science', *Harvests* 55/2 (2002), pp.13–14

27. Steiner, Rudolf, *Spiritual Foundations for the Renewal of Agriculture* (Bio-Dynamic Farming and Gardening Association, Inc. USA, 1993), trans. by C. Creeger and M. Gardner, p.135

28. Steiner, Rudolf, *Spiritual Foundations for the Renewal of Agriculture* (Bio-Dynamic Farming and Gardening Association, Inc. USA, 1993), trans. by C. Creeger and M. Gardner, p.20

29. See Gardner, Malcolm, 'When is Venus behind the Sun?', *Applied Biodynamics* 23/1998; and La Rooij, Marinus, 'The Astronomy of Peppering', *New Zealand Bio-Dynamic Farming & Gardening Association Newsletter* 47/4 (1994), p.27

30. See Moodie, Mark, 'Biodynamic peppers for pest control', *Star & Furrow* 115 (Summer 2011), p.49

31. See Bouchet, François, *L'Agriculture Bio-Dynamique* (Deux Versants, Paris, 2003), p.168

32. Pfeiffer, Ehrenfried, *Weeds and What They Tell* (Biodynamic Farming & Gardening Association USA, 1970), pp.10–11.

33. Masson, Pierre, *A Biodynamic Manual* (2nd edition 2014, Floris), p.181

34. Thun, Maria, *Results from the Sowing and Planting Calendar* (Floris, 2003) trans. by G. Staudenmaier, p.203.

35. Steiner, Rudolf, *Spiritual Foundations for the Renewal of Agriculture* (Bio-Dynamic Farming and Gardening Association, Inc. USA, 1993), trans. by C. Creeger and M. Gardner, p.65

36.   This is one reason why compost piles are sited above ground, ideally. See Chapter 3, siting compost piles.

37.   Steiner, Rudolf, *Spiritual Foundations for the Renewal of Agriculture* (Bio-Dynamic Farming and Gardening Association, Inc. USA, 1993), trans. by C. Creeger and M. Gardner, pp.91–93.

38.   Proctor, Peter, with Gillian Cole, *Grasp the Nettle* (Random House New Zealand, 1997), p.102

39.   Masson, Pierre,. *A Biodynamic Manual (2nd* edition 2014, Floris), p.93

40.   Bouchet, François, *L'Agriculture Bio-Dynamique* (Deux Versants, Paris, 2003), p.163

41.   Bouchet, François, *L'Agriculture Bio-Dynamique* (Deux Versants, Paris, 2003), pp.120–21

42.   Masson, Pierre, *Guide pratique de la bio-dynamie* (France, 1998), p.33

CHAPTER 6

1.   Masson, Pierre, *A Biodynamic Manual* (2nd edition 2014, Floris), p.112

2.   Masson, Pierre, *A Biodynamic Manual* (2nd edition 2014, Floris), p.150

3.   Bouchet, François, *L'Agriculture Bio-Dynamique* (Deux Versants, Paris, 2003), p.162

4.   Proctor, Peter, with Gillian Cole, *Grasp the Nettle* (Random House New Zealand, 1997), p.85 says spraying valerian 507 on grain improves its germination. Hugh Courtney in 'The Michaelmas Preparation: BD#504 Stinging Nettle', *Applied Biodynamics* 24/1998, p.6 says stinging nettle 504 seed baths benefit barley.

5.   Masson, Pierre, *A Biodynamic Manual* (2nd edition 2014, Floris), p.111

6.   von Wistinghausen, C., Scheibe, W., von Wistinghausen, E. and König, U., *The biodynamic spray and compost preparations production methods* Booklet 1 (Biodynamic Agricultural Association UK, 2000), p.38

7.   Thun, Maria, *Results from the Sowing and Planting Calendar* (Floris, 2003), trans. by G. Staudenmaier, p.175

8.   Thun, Maria, *Results from the Sowing and Planting Calendar* (Floris, 2003), trans. by G. Staudenmaier, p.175

9.   Masson, Pierre, *A Biodynamic Manual* (2nd edition 2014, Floris), p.130

10.   Masson, Pierre, *A Biodynamic Manual* (2nd edition 2014, Floris), p.112

11.   Masson, Pierre, *A Biodynamic Manual* (2nd edition 2014, Floris), p.121

12.   Thun, Maria, *Results from the Sowing and Planting Calendar* (Floris, 2003), trans. by G. Staudenmaier, p.130

13. Sattler, Friedrich and von Wistinghausen, Eckard, *Bio-Dynamic Farming Practice* (Bio-Dynamic Agricultural Association UK, 1992) trans. by A. Meuss, p.78

14. Lorand, Andrew, 'Backyard Biodynamics', *Biodynamics* 254/2005, p.21

15. Thun, Maria, *Results from the Sowing and Planting Calendar* (Floris, 2003), trans. by G. Staudenmaier, p.175

16. Proctor, Peter, 'Making liquid seaweed fertiliser', in *Biodynamic Perspectives*, ed. G. Henderson (New Zealand Biodynamic Association, 2001), p.13

17. Courtney, Hugh, 'The Michaelmas Preparation: BD#504 Stinging Nettle', *Applied Biodynamics* 24/1998, p.7

18. Masson, Pierre, *A Biodynamic Manual* (2nd edition 2014, Floris), p.190

19. Masson, Pierre, *A Biodynamic Manual* (2nd edition 2014, Floris), p.190

20. Thun, Maria, *Results from the Sowing and Planting Calendar* (Floris, 2003), trans. by G. Staudenmaier, pp.175–177

21. Lovel, Hugh, *A Biodynamic Farm* (Acres USA, 2000), p.113

22. Bouchet, François, *L'Agriculture Bio-Dynamique* (Deux Versants, Paris, 2003), p.162

23. Masson, Pierre, *A Biodynamic Manual* (2nd edition 2014, Floris), p.127

24. Masson, Pierre, *Guide pratique de la bio-dynamie* (France, 1998), p.27

25. Masson, Pierre, *Guide pratique de la bio-dynamie* (France, 1998), p.28

26. See Ingham, Dr Elaine, 'Compost tea, promises and practicalities', *Acres USA* 33/12 (Dec 2003)

27. von Wistinghausen, C., Scheibe, W., von Wistinghausen, E., and König, U., *The biodynamic spray and compost preparations production methods* Booklet 1 (Biodynamic Agricultural Association UK, 2000), pp.50–51

28. Masson, Pierre, *A Biodynamic Manual* (2nd edition 2014, Floris), p.129

29. Masson, Pierre, *Guide pratique de la bio-dynamie* (France, 1998), p.27

30. Masson, Pierre, *A Biodynamic Manual* (2nd edition 2014, Floris), p.113

31. Bouchet, François, *L'Agriculture Bio-Dynamique* (Deux Versants, Paris, 2003), p.156

32. Masson, Pierre, *A Biodynamic Manual*, (2nd edition 2014, Floris), p.137

33. *Lithotamnium calcareum* can also be sprayed directly on vines as a powder or diluted in herb teas made from stinging nettle or common horsetail 508, but not with chamomile, however. It is sprayed from flowering to bunch closure to help strengthen leaf and berry tissue,

stimulate chlorophyll formation and heal wounds caused by larvae or hail and, at bunch closure, to reduce humidity and the risk of subsequent grey (*botrytis cinerea*) and white rot (*penicillium* spp.).

34. Bouchet, François, *L'Agriculture Bio-Dynamique* (Deux Versants, Paris, 2003), p.132

35. Proctor, Peter with Gillian Cole, *Grasp the Nettle* (Random House New Zealand, 1997), p.45

36. Sattler, Friedrich and von Wistinghausen, Eckard, *Bio-Dynamic Farming Practice* (Bio-Dynamic Agricultural Association UK, 1992) trans. by A. Meuss, p.77

37. Courtney, Hugh, *Applied Biodynamics* 24/1998, p.5

38. Masson, Pierre, *A Biodynamic Manual* (2nd edition 2014, Floris), p.151

39. Proctor, Peter with Gillian Cole, *Grasp the Nettle* (Random House New Zealand, 1997), p.68

40. Masson, Pierre, *A Biodynamic Manual* (2nd edition 2014, Floris), p.168

CHAPTER 7

1. Steiner, Rudolf, *Spiritual Foundations for the Renewal of Agriculture* (Bio-Dynamic Farming and Gardening Association USA, Inc. 1993), trans. by C. Creeger and M. Gardner, p.20

2. Soper, John, *Bio-Dynamic Gardening* (Souvenir Press, 1996), eds. B. Saunders-Davies and K. Castelliz, pp.10–11

3. The IAU actually demarcated thirteen rather than twelve zodiacal constellations along the ecliptic by including Ophiuchus or Snake Holder which lies between Scorpion and Archer, but this has never featured in celestial biodynamic literature to my knowledge.

4. See Kollerstrom, Nick, *Gardening & Planting by the Moon 2004* (Foulsham, 2003), p.6

5. Ayurveda also works with the four elements plus a fifth, called the ether.

6. See Wachsmuth, Gunther, *The Etheric Formative Forces in Cosmos, Earth, and Man* (Anthroposophic Press USA, 1932); and Marti, Ernst., *The Four Ethers: Contributions to Rudolf Steiner's Science of the Ethers, Elements-Ethers-Formative Forces* (UK, 1984).

7. Spiess, Hartmut, '*Chronobiologische Untersuchungen mit besonderer Berücksichtigung lunarer Rhythmen im biologisch-dynamischen Pflanzenbau,*' Habil. Schrift Witzenhausen. Schriftenreihe, Institut für Biologisch-Dynamische Forschung, Darmstadt

8. See Nick Kollerstrom's letter in *Star & Furrow* 100/2009, p.40

9.   Steiner, Rudolf, *From Beetroot to Buddhism, Answers to Questions, 16 discussions with workers, Dornach, March 1–June 25, 1924* (Rudolf Steiner Press, 1999), trans. by A. Meuss

10. Proctor, Peter with Gillian Cole, *Grasp the Nettle* (Random House New Zealand, 1997), pp.114–5

11. Joly, Nicolas, *What is Biodynamic Wine?* (Clairview, 2007), trans. by M. Barton, pp.96–100

12. Dates for spring planting would have included 2/3 and 30/31 March 2007, 25 to 27 April 2007, 23 to 25 May 2007, 18 to 20 March 2008, 15 April 2008 (a short period due to a node), 12 to 14 May 2008, 8 to 10 March 2009, 5/6 April 2009 and 2/3 May 2009.

CHAPTER 8

1.   *Demeter processing standards for the use of Demeter Biodynamic® and related trademarks June 2015* (Demeter-International e.V.), p.73

2.   Steiner, Rudolf, *Spiritual Foundations for the Renewal of Agriculture* (Bio-Dynamic Farming and Gardening Association, Inc. USA, 1993), trans. by C. Creeger and M. Gardner, p.27

3.   *Demeter production standards for the use of Demeter Biodynamic® and related trademarks June 2015* (Demeter-International e.V.), p.18

4.   *Demeter production standards for the use of Demeter Biodynamic® and related trademarks June 2015* (Demeter-International e.V.).

5.   See Costello, Michael J., and Daane, Kent M., 'Arthropods' (Chapter 8), in Ingels, C., Bugg, R., McGourty, G., and Peter Christensen, L., *Cover cropping in vineyards* (University of California publication 3338, 1998), p.94

6.   The EU (1992–2011) used the word 'from' and the US 'with' when referring to wine made with/from organic grapes/viticulture.

7.   See *Demeter processing standards for the use of Demeter, Biodynamic® and related trademarks as at June 2015 – to be implemented by each member country by the 1ˢᵗ July 2016* (Demeter-International e.V.), pp.72–78

8.   Joly, Nicolas, *What is Biodynamic Wine?* (Clairview, 2007), trans. by M. Barton, p.46

9.   Bouchet, François, *L'Agriculture Bio-Dynamique* (Deux Versants, Paris, 2003), p.152

INDEX